Fran Balkwill and Mic Rolph
with Victor Darley-Usmar

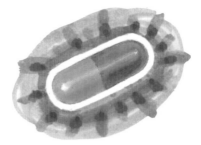

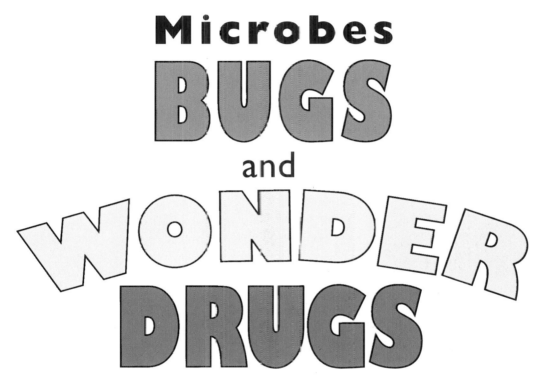

Microbes
BUGS
and
WONDER
DRUGS

Potions to Penicillin, Aspirin to Addiction

PORTLAND PRESS

Making **sense** *of* **science**

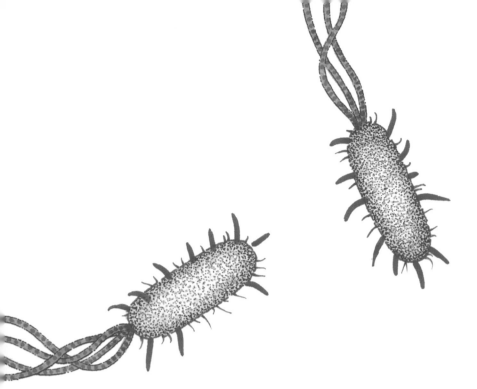

First published in 1995 by Portland Press Ltd.
59 Portland Place, London W1N 3AJ

Copyright © text Fran Balkwill 1995
Copyright © illustrations Mic Rolph 1995

The authors assert the moral right to be identified
as the authors of the work

ISBN 1 85578 065 8   ISSN 1355 8560

Printed by Cambridge University Press, Cambridge, U.K.

Contents

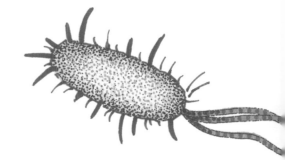

Note to the reader:
If you are not familiar with any of the scientific terms we use in this book, turn to the Glossary at the end. It should give you more information and tell you how to pronounce the words.

Nothing's New

Since the dawn of human history, disease has stalked our species. Driven by their animal instincts, prehistoric humans must have quickly learned what could be safely eaten and what made them feel strange or ill. They discovered plants, berries, tree barks and minerals that could heal wounds and make the sick feel well again. It soon became a special skill to understand natural medicines, and collect or grow the herbal remedies.

In every ancient civilization, medicine men, witch doctors, shamans and wise women were all-powerful in treating disease. They dispensed their medicines with ritual, magic and religion. They learned to strengthen their potions by making pastes or liquid extracts. By trial and error, the early medicine men and women found treatments for most afflictions. Recipes were passed from generation to generation, without an exact understanding of how they worked. These precious medicines could relieve pain, stop fits, sedate, stimulate, purge or soothe. Some could also intoxicate and cause very weird hallucinations.

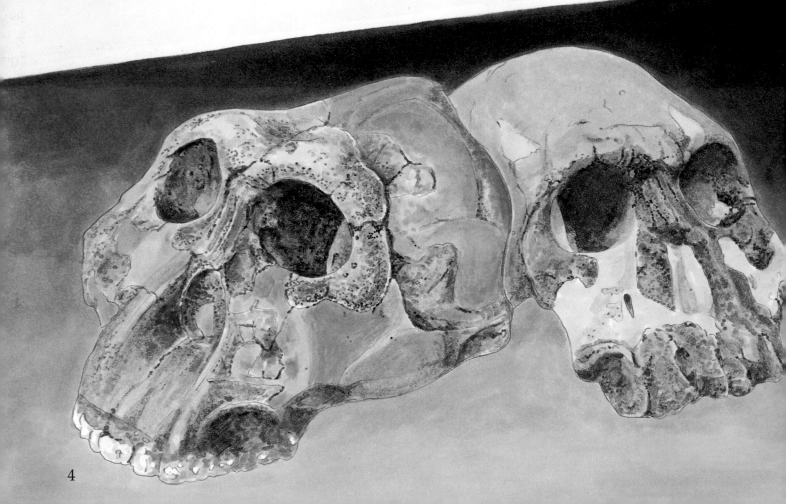

But in ancient times, many people died before the age of 30 or 40. In spite of the power of primitive medicine, generations were ravaged by disease, as well as famine, war and natural disasters.

During the last 150 years, scientists and doctors have begun to understand how the first medicines worked. They have learned that the power of the crude potions came from special chemicals, called drugs, which could be extracted from the herbal remedies or made in a laboratory.

As scientists have unravelled the intricate workings of the natural world, many new drugs have been discovered. Modern medicines are an important part of your life, and they are usually more effective than ancient potions. Modern drugs may even save your life, or (just possibly) destroy it.

But what exactly are drugs?

they give feelings without emotions

they're dangerous!

they are remedies

CRACK!

PILLS

You can drink them
swallow them
smoke them
and snort them

they make you better

chemicals that screw
up your head!

ACID

painkillers

I'm not really sure!

they help you

if you take too many
they make you ill

COCAINE

You can inject them

people get money
for selling them

you take them for toothache

HEROIN

Drugs are awful

they get rid of your pain

DRUGS ARE
POTIONS

stimulants

GRASS

I've never ever
thought about them

Mindbenders

some are bad

6

I wouldn't really know!

POT

Booze?

antibiotics

they let you rave all night!

Tablets your Doctor gives you

GANGA

CHEMICALS THAT ALTER YOUR MOOD

they relieve your problems

I wouldn't know

CIGARETTES?

Tablets?

Horrible!

they're terrible!

ECSTASY!

YOU CAN GET THEM FROM A CHEMIST

you can become addicted to them

HASH

you get them from pushers!

DOPE!

some are good

Cannabis

WHIZZ!

Are they medicines?

They help you get better when you're ill!

The word 'drug' means different things to different people. In fact, a drug is any chemical that can change the way a living creature functions. Drugs can be simple chemicals or complicated ones. They may be gas, liquid or solid. Drugs can be inhaled, swallowed, absorbed through the body surface, or injected. But all drugs work in the same way. They get very close to natural chemicals in living cells and alter what they do.

You might or might not know that all chemicals are made from building blocks called molecules and that a molecule is a mixture of atoms that are joined together in a particular way.

A molecule of water, for instance, is made from two atoms of hydrogen and one atom of oxygen.

One of the simplest and oldest drugs is alcohol. It is made of two atoms of carbon, six atoms of hydrogen and one of oxygen.

The most important thing about drug molecules is their shape. Using complicated computer programs scientists can work out these shapes. This drawing shows a molecule of alcohol. Two carbon atoms are shown in black, the oxygen atom is red and green represents hydrogen.

One other thing to remember about molecules is that they are very, very small. A pin head, for instance, is about 10 million times wider than a molecule of water. There are over 30 million, million, million, million molecules in each litre of water.

Most drugs work because they bind tightly to molecules inside, or on the surface of, your cells. Once a drug molecule is stuck to a particular cell molecule, it interferes with the normal working of that molecule. And the reason why a drug works is because of its shape. Each drug has a particular shape that allows it to bind tightly to one type of molecule in your cells.

Most of the drugs we are going to tell you about bind to proteins. Proteins are important molecules that make your cells the size and shape they are. They allow cells to move and communicate with each other. Proteins called enzymes help the cell carry out chemical reactions. Proteins control the transport of substances in and out of the cell.

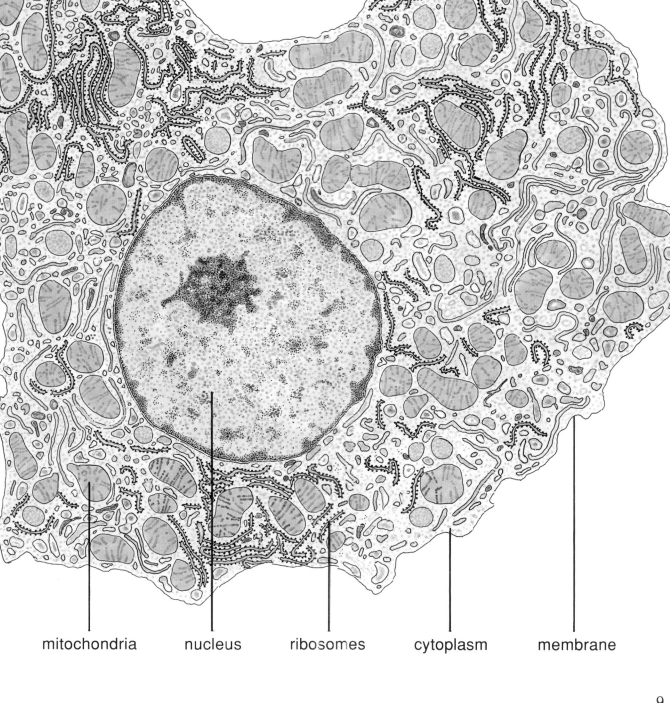

mitochondria   nucleus   ribosomes   cytoplasm   membrane

Drugs that inter-
fere with the way
proteins work can
therefore be very
powerful. Other
drugs bind to the
genetic material in
cells, the DNA
blueprint that
makes each human
being unique and
controls the way
each cell behaves.
DNA is all-
powerful in the cell
because it contains,
in a chemical code,
recipes for all the
proteins that a cell
has to make. DNA
directs the cells to
make these proteins
as and when they
are needed. DNA is
a long thread-like
molecule found in
the nucleus of cells
in strands called
chromosomes.
If you unravel the
DNA thread from
one of the
chromosomes and
magnify it about 35
million times, you
can see that not one,
but two, strands
make up the thread.
They wind around
each other so that
the DNA is like a
twisting, twirling
ladder.

This shape is called
a double helix. Each
time a cell divides
to become two cells,
the DNA
instructions are
copied. The double
helix unzips so that
there are two single
strands of DNA.
Then each single
strand becomes a
pattern for another
strand. The
chemicals that
make DNA are
floating around in
the cell and join up
in a precise order.
The four chemicals
that make up the
DNA thread are
called **A**denine (red
in this drawing),
Thymine (green),
Cytosine (yellow)
and **G**uanine (blue).
Adenine always
joins up with
Thymine. **G**uanine
always joins up
with **C**ytosine. This
means that two
identical double
helices are made
when DNA is
copied. Drugs that
bind to DNA will
usually stop a cell
multiplying, and
are often lethal to
the cell.

10

Sometimes you can easily understand how a drug works when you look at a computer model of its shape. For instance, there is a drug that stops certain viruses multiplying in cells. This drug, acyclovir, is almost the same shape as guanine, one of the building blocks of the DNA genetic code. However, a particular type of virus is able to (mistakenly) use acyclovir instead of guanine to make its DNA. Once the virus has included the drug molecule in its DNA, the virus DNA strand cannot get any longer.

No more viruses are made. The virus infection is defeated! This is the computer model of the acyclovir molecule and the part of DNA that it resembles. The carbon molecules are coloured black, hydrogen green, oxygen red, and nitrogen blue.

Some of the most remarkable and important stories of drugs come from our desperate fight against the ravages of infectious disease, a fight that we seem to be winning for the time being, particularly against bacteria.

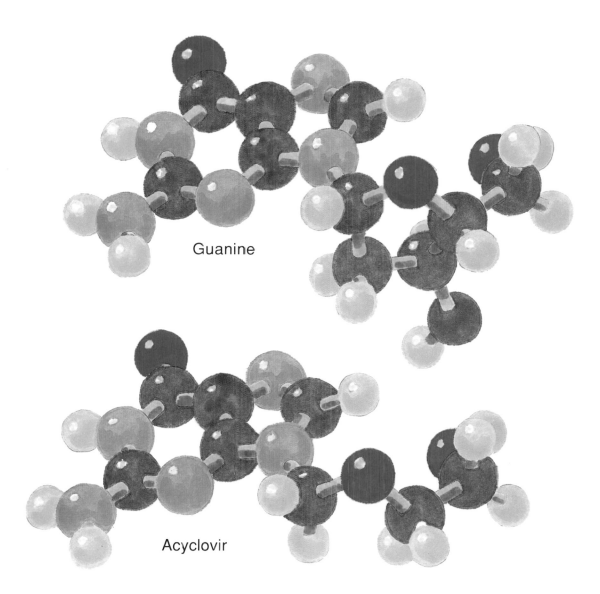

Guanine

Acyclovir

11

Microbes Bugs
and
Wonder Drugs

Bacteria are microbes, minute living organisms that can only be seen with a microscope. They have adapted to live in every possible habitat of our planet, where they help maintain the balance of living creatures and chemicals. Bacteria in the soil break down waste and trap nitrogen from air. Bacteria in the rivers, lakes and oceans are a crucial part of the food chain. You are personally dependent on billions of bacteria that peacefully inhabit your intestines! Not only do they help digestion of tough vegetables, but they manufacture B and K vitamins that are necessary for efficient body functioning.

Bacteria are about 5 to 50 times smaller than human cells, and they are quite different from your cells both inside and out. Surrounding and protecting the fragile bacterial cell membrane there is a tough cell wall which gives the bacterium its shape. Inside the membrane, the cytoplasm is a thick, almost transparent liquid containing all the usual chemicals that cells have. But there are no complicated structures such as you would find in human cells, apart from ribosomes, which are where proteins are made. The genetic material, DNA, in human cells is divided into 46 threads called chromosomes, but the DNA plans of bacteria are in one single circular chromosome thread, attached to the delicate membrane.

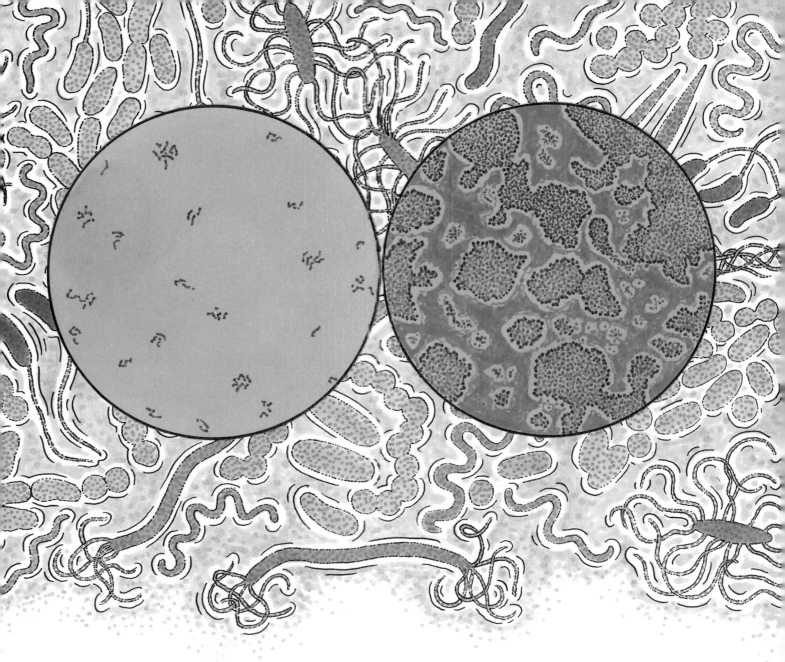

Although you cannot see individual bacteria with the naked eye, you can see them when they grow as colonies. A single bacterium, when it lands on a nutritious surface, will quickly multiply into millions of bacteria and form a glistening blob a few millimetres wide. A single generation of bacteria lasts as little as 20 minutes. In fact, if there was enough food available, one bacterium could multiply into a lump of bacteria larger than the entire planet, in just three days!

In spite of this, only a few kinds of bacteria prove to be harmful. If these microbes get inside your body, they can multiply rapidly and release poisons. Normally your defender cells that patrol your body to fight germs, attack quickly and successfully. But, occasionally, there are just too many bacteria, and you will become ill. Scientists and doctors now understand a lot about bacteria and how to prevent and cure the diseases they cause. Not so long ago, however, millions of humans died of devastating bacterial diseases that can be easily cured or prevented nowadays, and much of their food was destroyed by these invisible microbes.

In the early 1900s, hospital wards were filled with children and young adults dying from infections of the blood, lungs, skin and bones. Soldiers frequently died from infected wounds. The germs that caused this suffering had been discovered only 30 years before, germs called BACTERIA.

London 1928 Alexander Fleming was studying some of these dangerous bacteria. They grew as glistening colonies, each from a single bacterium that had multiplied into millions.

When he went on holiday, he left some bacteria growing in plates on the laboratory bench.

On his return, he saw nothing interesting, so he discarded the plates into the antiseptic tray.

Thankfully, not all the plates were submerged. Later, talking to a friend, he noticed something.

THAT'S FUNNY!

Where a strange blue/green mould contaminated the edge of one plate, no bacteria had grown. Further away the colonies were strangely pale. The mould made something that killed bacteria!

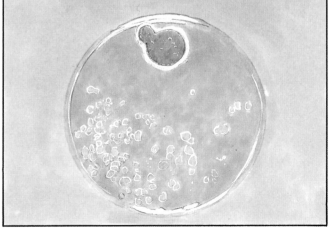

Fleming grew the mould, and made a liquid extract. The diluted 'mould juice' killed many types of dangerous bacteria. He named his discovery 'penicillin' and described it in a scientific paper.

ON THE ANTIBACTERIAL ACTION OF CULTURES OF A PENICILLIUM, WITH SPECIAL REFERENCE TO THEIR USE IN THE ISOLATION OF *B. INFLUENZÆ.*

ALEXANDER FLEMING, F.R.C.S.

From the Laboratories of the Inoculation Department, St Mary's Hospital, London.

Received for publication May 10th. 1929.

WHILE working with staphylococcus variants a number of culture-plates were set aside on the laboratory bench and examined from time to time. In the examinations these plates were necessarily exposed to the air and they became contaminated with various micro-organisms. It was noticed that around a large colony of a contaminating mould the staphylococcus colonies became transparent and were obviously undergoing lysis (see Fig. 1).

Subcultures of this mould were made and experiments conducted with a

There was little interest in Fleming's experiments. Purification of the mould juice was very difficult.

Crude penicillin lotion was used in wound and eye infections. There was little effect.

Oxford 1938 Howard Florey and Ernst Chain decided to do some more experiments with penicillin.

Paper! Paper!!

They made slow progress. It was a difficult and unusual molecule to work with, but an interesting challenge.

1939 Britain at war Mass evacuations, blackouts, rationing, and air-raid shelters changed a way of life. No one realized that experiments with a mould would help the war effort!

BRITI COMPLETE MOBIL MOBILISATION EVEN 8.30 N -OFFICIAL THE STAR

March 1940 At last Chain had 1/10th gram of impure powder. It did not harm mice. Now for Florey's crucial experiments!

May 1940 8 mice were given lethal doses of bacteria. One hour later 4 received a penicillin injection.

The results were clear. By the following morning, all the untreated mice had died, but penicillin saved the other 4 mice!

CRUDE PENICILLIN March 22ND 4 [SODIUM SALT]

Experiments with more mice and other bacteria convinced Florey and Chain that penicillin was a very powerful drug. Now from mouse to man!

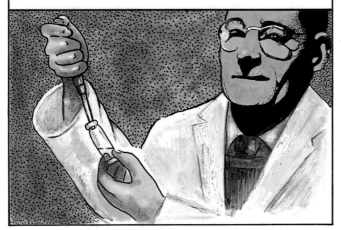

They improvised with bedpans, milk churns, and other dairy equipment. 40,000 litres of mould juice would be needed for five or six patients. The chemicals used in purification were enough to blow up the building. London was suffering the Blitz – who cared about the risk?

3000 times more penicillin was needed to treat each human patient. In war time, drug companies could not help. So Florey turned his laboratory into a makeshift factory.

Feb 1941 The first patient was a policeman dying of overwhelming infection. Within 24 hours he was dramatically better. Three more dying patients were cured, and two others improved for a while.

But supplies soon ran out. Sadly, the policeman got worse and died. But Florey already knew he had a 'wonder drug' – if he could make enough!

The answer lay in America. With equipment used for brewing beer, and a more powerful strain of the mould found growing on a melon, larger quantities of penicillin could be made. Companies in Britain also began scaling up production. By the Normandy Invasion in 1944, there was enough penicillin to treat all the casualties. Penicillin was also released for civilian cases. It became much cheaper to make and buy.

Normandy 1944 Medical officers were astonished by the lack of infections in soldiers. In hospitals around the world dying patients were dramatically cured.

World supplies of penicillin

June 1943	170	patients
June 1944	40,000	patients
June 1945	250,000	patients

Stockholm Sweden 1945

Fleming, Chain and Florey received the Nobel Prize for Medicine.

The age of antibiotics had arrived!

The idea that moulds might cure infections was not new. Some ancient Egyptian medicines included mouldy bread, and scientists in the 18th century knew that substances made by bacteria and moulds could kill other microbes. But when extracted, these antibiotics were rather weak and often poisonous. That is probably why there was so little interest when Fleming discovered penicillin. As the penicillin story unfolded, scientists realized that there might be other powerful and non-poisonous antibiotics in the world around them, particularly in the soil. Antibiotics were soon discovered that would work in infections where penicillin did not. Bacteria called *Streptomyces* were one useful source.

Some time before antibiotics were developed, scientists had an idea that anti-bacterial drugs could be made from textile dyes. This was because some of these dyes would colour bacterial, but not human, cells. They reasoned that such chemicals might also kill bacterial cells without making the patient feel too ill. These ideas led to a number of drugs, but few were as powerful as antibiotics.

Scientists began to alter antibiotics chemically so that they stayed in the body longer and were more resistant to stomach acid. Over 2000 different antibiotics have been developed since penicillin was discovered. Some of the latest ones will kill more than 98% of all the different types of bacteria isolated from hospital patients.

Have you ever wondered how antibiotics work?

Bacterial cells and human cells are different from each other in a number of ways. This is of course why most antibiotics kill bacteria and not humans! For instance:

Bacterial cells have strong outside walls, made of proteins and sugars, making a tough barrier, rather like 3-dimensional wire mesh. The shape of the penicillin molecule is very similar to one of the molecules in this mesh. As each bacterium grows, it makes more of its wall. Along with the usual molecules, bacteria use the penicillin molecules. Once penicillin is part of the wall, the links of the mesh are not strengthened because penicillin cannot be bound to the other molecules. The wall is weakened. As the bacteria try to divide, they burst apart. Human cells don't have walls, so they are not affected.

Ribosomes that make proteins in bacteria are different from most human ribosomes. Some antibiotics stop bacterial ribosomes working. This means that the bacteria can't make proteins properly and they die. There is one problem, though. The energy factories of human cells, called mitochondria, have their own ribosomes, and, unfortunately, these are like the bacterial ones. So human cells are sometimes harmed by antibiotics although the damage is usually not serious. The membrane inside the bacterial cell wall is another target for antibiotics.

Not all antibiotics kill bacteria, some just stop them growing. Bacteria that can't multiply are a sitting target for defender cells that patrol the blood and tissues and fight invading organisms.

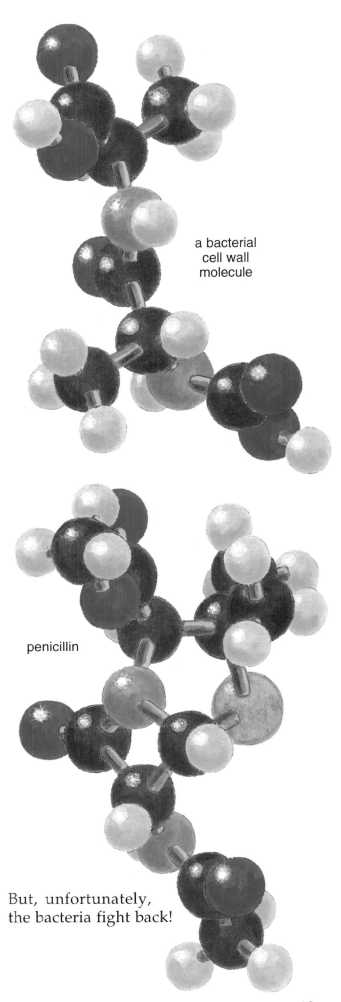

a bacterial
cell wall
molecule

penicillin

But, unfortunately,
the bacteria fight back!

Bacteria can change so that antibiotics can't destroy them, and all because of DNA. Sometimes a bacterium gets a mutation in its DNA genetic instructions. This means that it makes an altered protein. This can sometimes alter the bacteria so that antibiotics cannot kill it. All offspring of that mutant bacteria will be resistant to the antibiotic. But worse than this, resistance to the antibiotic can be transferred to other bacteria nearby. This is because bacteria often exchange DNA. Some bacteria release their DNA, particularly when they die. This DNA is taken up by different bacteria nearby and can become part of their chromosomes. Bacteria even pass bits of DNA from living cell to living cell. They join together, making a temporary tube through which tiny bits of DNA called plasmids can pass. Some plasmids contain instructions for as many as six altered proteins, enough to make bacteria resistant to six different drugs.

Bacteria can make enzymes that chop up antibiotics and render them useless. For instance, some bacteria make an enzyme called penicillinase. This makes them resistant to penicillin. Over the years, chemists have altered the penicillin molecule to make new antibiotics that cannot be destroyed by penicillinase. They have also made changes so that the antibiotic can kill more different types of bacteria. This has been done so many times over the last 40 years that there are now hundreds of different types of penicillin, all slightly different in their shape.

But the bacteria still fight back! All this means that we must use antibiotics sparingly and only when it is really essential. New antibiotics will always be needed, because bacteria will never run out of ways of becoming resistant.

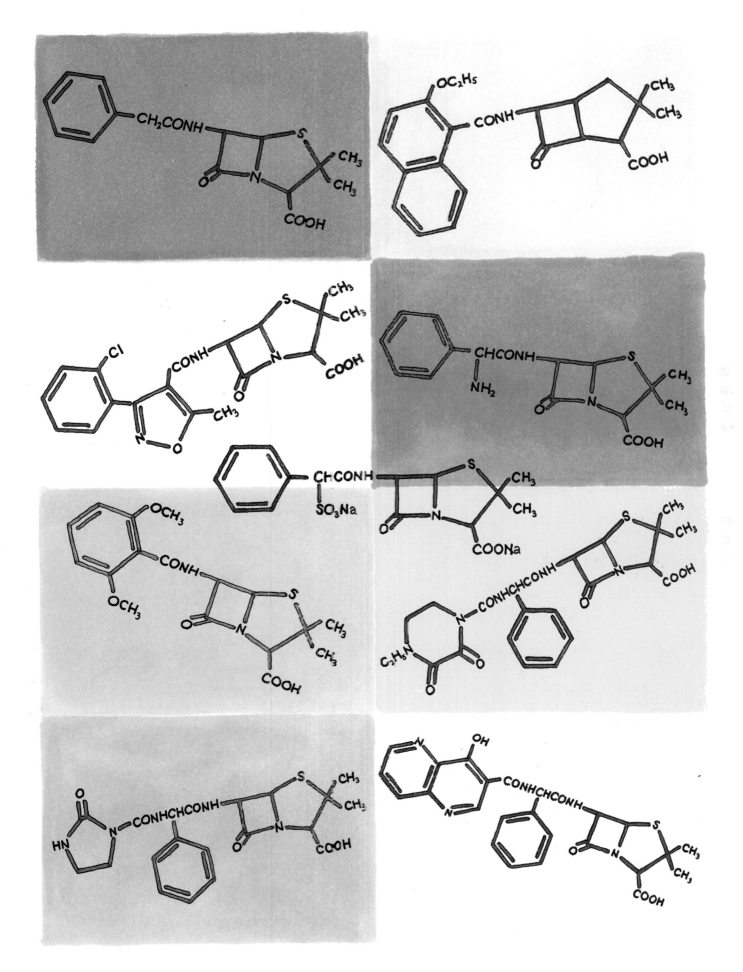

21

Some years after bacteria were first discovered, scientists began to suspect that there were other germs, much smaller than bacteria, and quite different in their behaviour. They called these mysterious agents **viruses** (*virus* is Latin for poison). Viruses are parasites. When outside a cell, be it human, plant or bacteria, they are lifeless particles made of nucleic acid, fats and proteins. But when they invade a cell, they become very much alive, taking over the cell's machinery to make many more viruses, usually destroying the cell in the process.

Inside each virus you will find genetic instructions. These genetic instructions are surrounded by a protective coat of proteins. In some types of virus, this coat is surrounded by a fatty membrane.

Some viruses contain DNA as their genetic material. They usually copy this in the nucleus of a cell and make their proteins in the cytoplasm.

Other viruses have RNA, not DNA, as their genetic instructions. RNA is the messenger strand of DNA that carries the DNA code to the protein-making part of the cell.

Some RNA-containing viruses simply copy their RNA and make proteins that way. But others do something quite extraordinary, they make DNA from RNA. The virus DNA then, somehow, becomes part of a human chromosome. From this hidden position, the virus may remain silent, for days, months, and even years, before it begins directing the manufacture of millions of new particles.

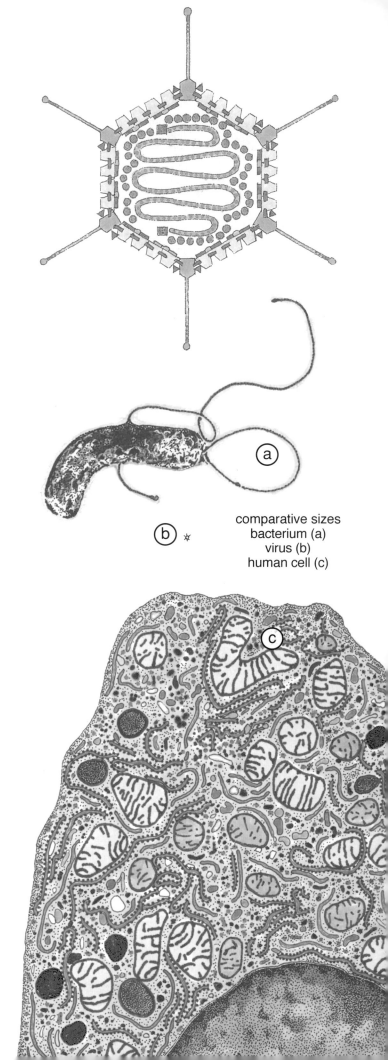

comparative sizes
bacterium (a)
virus (b)
human cell (c)

22

A virus containing DNA attaches to the membrane of a human cell.

As it penetrates the cell, the virus protein coat is destroyed.

The DNA of the virus travels to the nucleus. There it uses the cell's enzymes to make an RNA messenger strand.

The RNA messenger strand travels to the cytoplasm and makes virus proteins, which then pass back to the nucleus.

In the nucleus many copies of the virus DNA are made, and new virus particles assembled from DNA and protein.

The new viruses break out of the cell and infect more cells, which make more viruses to infect more cells...

This year, about 90% of the people in Great Britain (including you!) will suffer from some sort of viral infection. In fact over half the infections that afflict humans are caused by viruses. Although a virus infection can make you feel quite ill, your own defence system will eventually beat most viruses and you will soon be better.

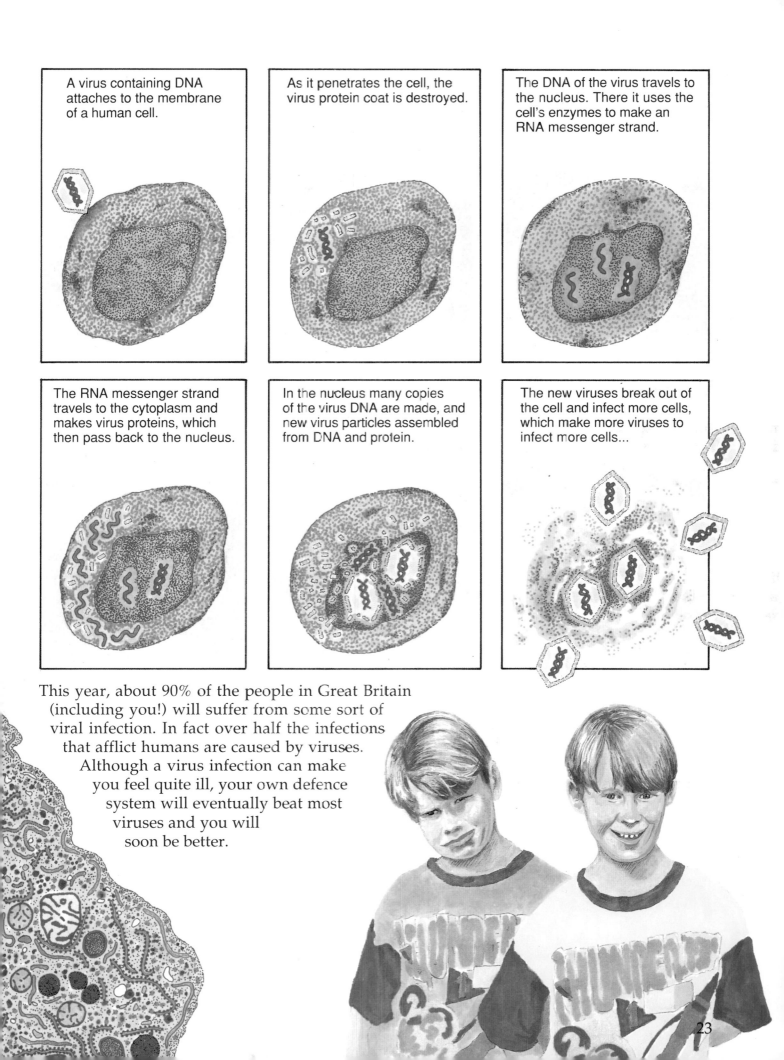

There are very few drugs that can treat virus infections. This is because viruses multiply only inside your cells, taking over many enzymes and structures. Anything that stops viruses growing is likely to harm human cells as well.

Acyclovir could stop viruses growing inside cells and stop them spreading to other cells. Unfortunately, it could only attack one type of virus, herpes virus. However, herpes viruses cause several common illnesses like chickenpox, shingles, cold sores and pneumonia in babies. They can also be very dangerous to people whose immune system is not working properly.

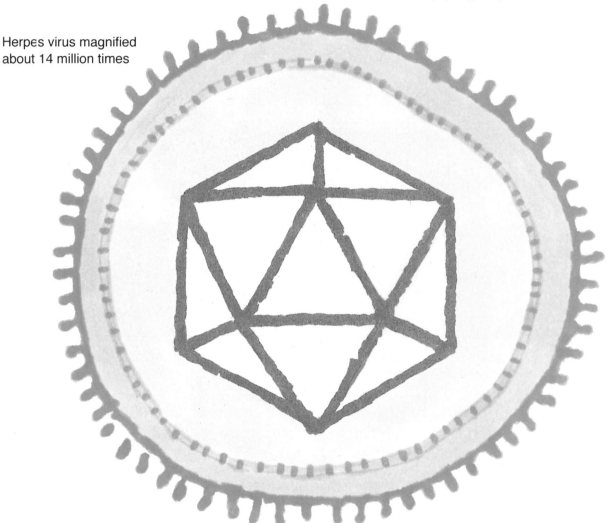

Herpes virus magnified about 14 million times

Between 1950 and 1980, scientists carried out many experiments on nucleic acids. They made many different compounds that interfered with DNA and RNA in cells, in the hope that some of these might be antiviral drugs. They finally hit lucky with a powerful compound called acyclovir. Acyclovir resembles guanine, one of the chemicals that make DNA and RNA.

One rather interesting fact about herpes viruses is that they lurk around your body, particularly in nerve cells, long after any signs of infection have gone. Many weeks, months or years after the original infection, they can start to multiply and cause disease once more.

24

This is how acyclovir works

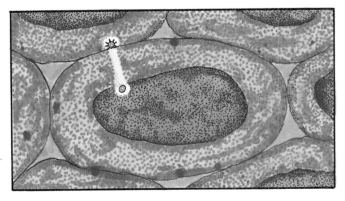

A herpes virus infects a cell by fusing with the cell membrane. The virus coat dissolves away and the virus DNA is transported to the nucleus of the cell.

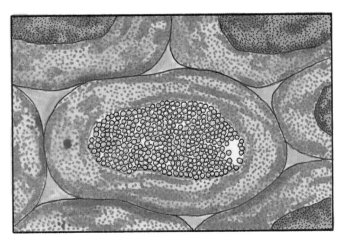

The virus DNA takes over the nucleus and makes thousands more viruses.

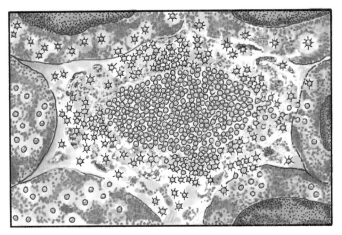

New virus particles are assembled, and bud out from the nucleus ready to attack cells nearby. The cell is doomed!

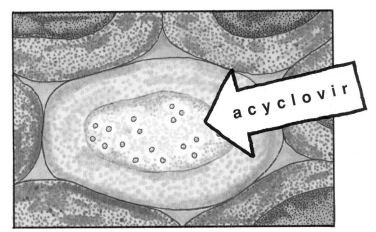

Acyclovir is able to pass through the cell membrane and get to the nucleus. The virus is tricked into using acyclovir instead of guanine to make its DNA.

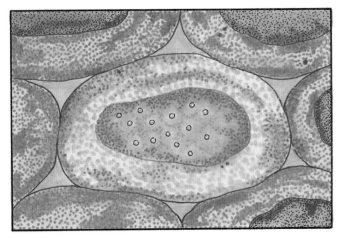

Once acyclovir is part of the virus DNA, the thread cannot be lengthened. No more DNA can be made and so virus production stops.

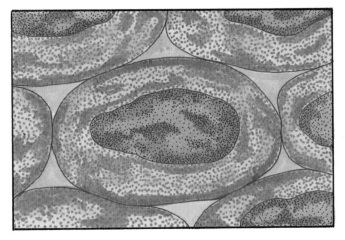

Human cells make their DNA in a slightly different way. Our cells can't use acyclovir so they are not harmed.

How herpes viruses cause trouble...

Cold sores are caused by herpes viruses. You may have caught this virus recently by skin contact or in the air...

...or you may have caught it years ago.

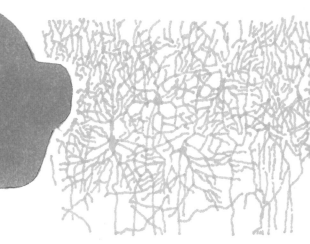

These viruses love to lurk in nests of nerve cells at the base of your neck. They can stay there without causing any trouble for years.

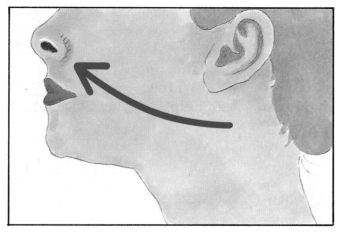

Suddenly they begin to multiply and spread down the nerve cell axons to cells near your mouth. The first feeling is often a tingling sensation.

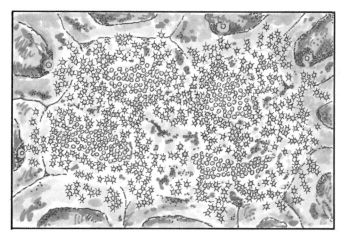

The infected cells swell uncontrollably. They eventually burst releasing millions of new viruses that infect the cells nearby.

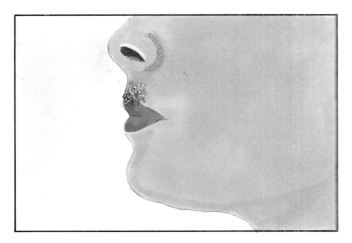

The fluid from the swollen cells forms blisters under the skin. The cold sore becomes red and itchy.

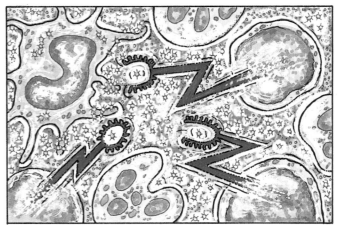

The immune system begins to fight back and limit the spread of this destructive virus. Virus-infected cells are destroyed by lymphocytes, and antibodies stick to viruses so they cannot infect cells any more.

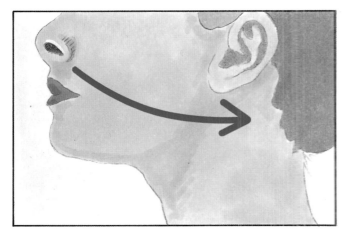

Within about 10 days, the virus will be beaten. Macrophages clear up the mess, skin cells repair the damage, and the virus retreats up the nerve cell axons.

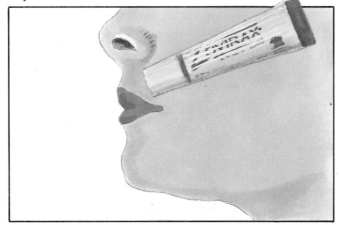

But if acyclovir cream is used at the first signs of skin tingling, it can penetrate the dead outside layer of the skin and get inside the cells that are in trouble.

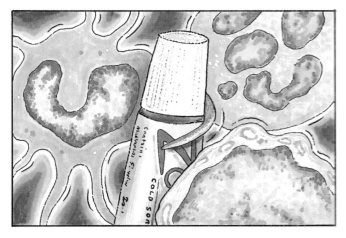

The viruses can no longer make DNA, so no more viruses are made. Acyclovir works with your defender cells to beat the virus in about half the usual time.

There are other rarer but more serious infections caused by herpes viruses, particularly infections of the brain and lung. Acyclovir, and a few similar drugs, can help patients with these infections. But drugs like acyclovir are the only really successful anti-viral drugs developed so far.

Another strategy for fighting viruses is to harness part of the body's own defence system, the interferons.

In 1957, scientists Alick Isaacs and Jean Lindemann discovered that cells infected with a virus fight back by making a protein that protects other cells from virus infection. They called the protein 'The Interferon'.

At first there was great excitement. Crude extracts of interferon powerfully protected cells from infection with many different types of virus. In the test tube, one million millionth of a gram of interferon would protect one million cells from attack by ten million viruses. Surely this was a 'wonder drug' for curing virus infections? But they soon found that there were many interferons (your body makes about 20!). And it was almost impossible to make enough interferon to treat more than a handful of patients. In the 1970s most of the world's supply of interferon was made in Finland, and it was an arduous and expensive task. For instance, in 1977 it took 14 scientists one year to make a quarter of a gram of interferon from the white blood cells in 90,000 blood donations.

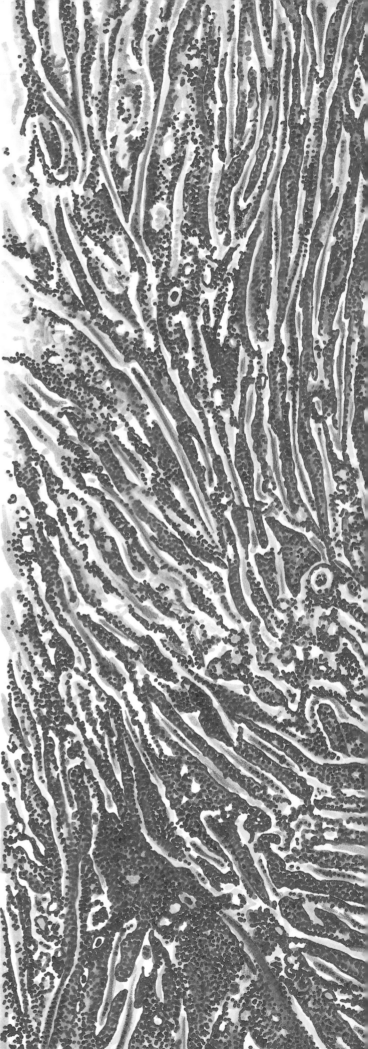

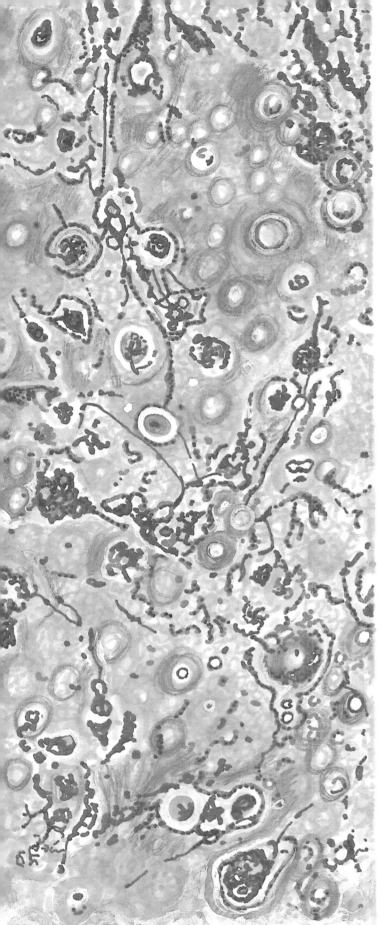

The cells in these two pictures have been infected with ten million virus particles. Those on the left were also given one million millionth of a gram of interferon. The virus killed the cells on the right, interferon protected those on the left.

Genetic engineering, and other scientific advances, changed all this. DNA from the interferon gene was inserted into bacteria. The bacteria became microscopic interferon factories. Small vats of easy-to-grow bacteria could produce enough interferon to treat hundreds of patients within a few hours! Equally pure interferons were also made from huge vats of human cells. At last there was plenty of interferon. Was interferon really a wonder drug for virus infections? Initially the answer was a disappointing 'no'. It seemed that people had to take interferon before they met the virus. Hundreds of people were given a squirt of interferon up their nostrils each morning to see if they were protected from the usual cold and 'flu viruses that attack each winter. Unfortunately, the interferon spray itself caused a runny nose and bleeding. Interferon injections were not without side-effects either. The patients felt as if they had 'flu. The injection gave them fever, tiredness and all kinds of aches and pains. (Scientists now think that virus infections make you feel ill, not just because some of your cells are being destroyed, but because your body is producing large quantities of substances like interferons, to fight the virus.)

But later on, doctors found that interferon could help people infected with a virus that attacks the liver. In some parts of the world millions of people suffer from chronic liver infections that never seem to get better. Interferon helps about half of the people with liver infections, and stops their livers being destroyed. It seems to work by boosting the body's natural defences against the viruses. But interferon does not seem to act against other chronic virus infections, particularly one that concerns us all, AIDS (Acquired Immunodeficiency Syndrome).

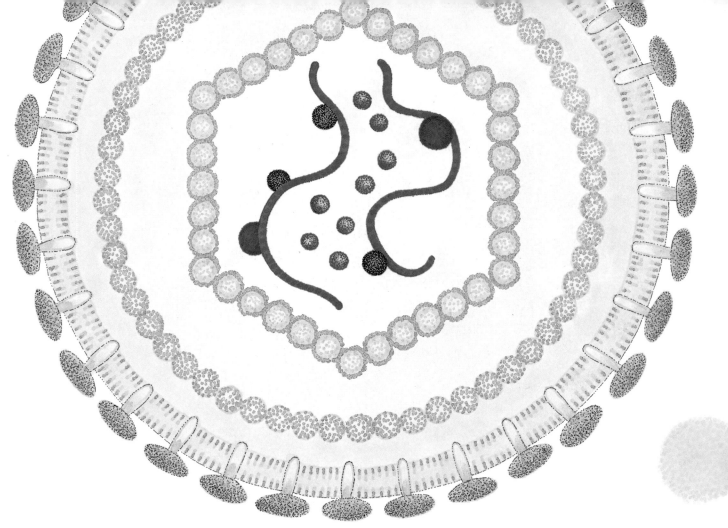

Throughout human history, new diseases have appeared to threaten our species. The most frightening disease of the late 20th century is AIDS, caused by the Human Immunodeficiency Virus, HIV, which probably came from a mutated virus of central African apes. The virus may have occurred in isolated areas for some time before its rapid spread. Migration, war and famine probably helped its progress. In 1981 the first cases of a strange new sickness were seen in Haiti and North America. Ordinarily fit and healthy young men were succumbing to unusual infections by microbes usually harmless to humans. It seemed as if their immune system was being destroyed. Most of these men were homosexual. The disease began to spread, first to drug abusers who shared syringes, to people who were given blood transfusions, or treated with blood products because they had a clotting disease (e.g. haemophilia).

Doctors and scientists realized that AIDS was spread by exchange of body fluids. Soon the most common means of infection was heterosexual sex. AIDS spread to Europe and Asia and back to Africa, affecting men and women in every section of society. Even babies of infected mothers succumbed to the disease. The HIV virus was isolated in 1983. As scientists began to understand its life cycle, they realized AIDS would be especially difficult to treat. HIV virus can only invade certain types of white blood cells and nerve cells. Its genetic information is a single strand of RNA. When it infects a cell, the virus turns its RNA into DNA. This becomes part of a chromosome. The cell now carries a deadly message that can lie hidden for many months or even years. When the HIV DNA becomes active, it begins to make many viruses. They bud out from the cell to infect more cells. The cell then dies and is not replaced.

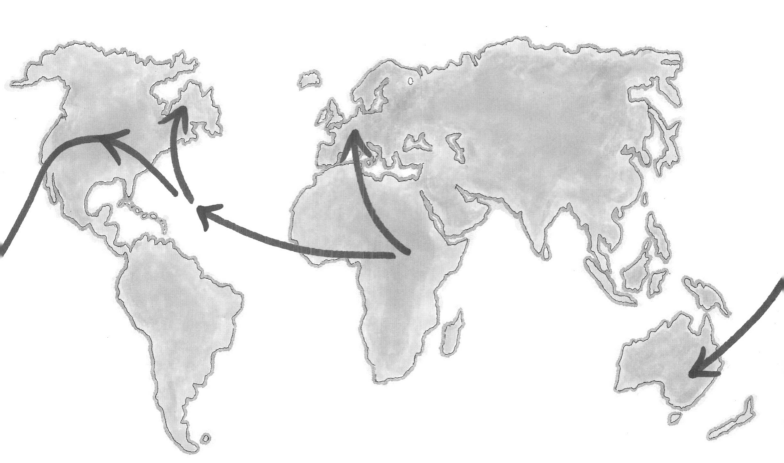

People who carry the AIDS virus can be healthy for many years. They even make antibodies and killer cells to fight the virus, but nothing can halt the silent destruction of their immune system. While this is going on, they can spread the infection. In 1981 AIDS was first discovered in 31 American men. It is predicted that by the year 2000 over 40 million adults will have been infected, the vast majority by heterosexual sex. Eight to ten million children will have caught the infection from their mothers, and another ten or so million orphaned because their mothers have died of AIDS. The worst places for the epidemic are Africa and Asia.

As you can imagine, it is very difficult to attack a virus when it has hidden its DNA inside a human chromosome.

A drug called azidothymidine, or AZT, was discovered in 1984. It stops the virus making DNA from its RNA, but doesn't seem to affect human DNA as much. It is quite toxic, but it helps some seriously ill patients, and may stop pregnant women transmitting the virus to their babies. New drugs are desperately needed. In a move that reflects the dangerous nature of the AIDS epidemic, 15 of the biggest drug companies in the world have decided to combine their knowledge and work together for a cure.

The good news is that HIV is not infectious like 'flu or measles. It is most commonly spread by unprotected sexual contact. So simple changes in behaviour, like using condoms, can halt the spread of the epidemic while scientists find ways of fighting it. Finding one simple, non-toxic pill may be very difficult, but...

...there is another way to defeat viruses and bacteria.
This treatment does not work against all the germs that attack our species, but it can be extremely successful. It protects us from some of the most dangerous diseases and may be able to rid the entire planet of our deadliest microscopic enemies, including the HIV virus. This treatment is called vaccination, and the story of smallpox is its greatest triumph (so far).

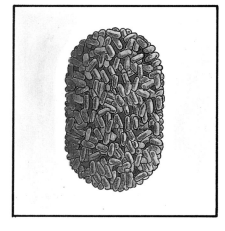

The smallpox virus used to cause one of the most frightering and destructive diseases ever known.

In societies where it thrived, most of the population would succumb. The survivors were often disfigured and blinded.

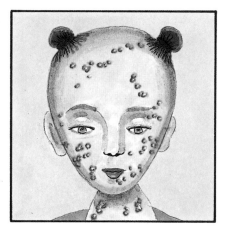

No one knew what caused this scourge, but once you had smallpox, however mild, you could not catch it again.

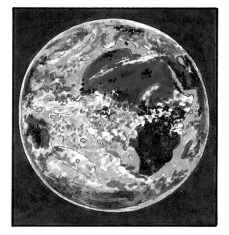

For many thousands of years people all over the world devised ways of 'safely' catching the disease, or *buying the pox*.

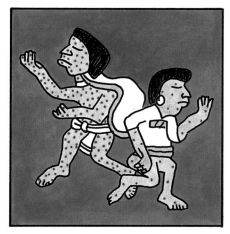

Young children, in particular, were put with people who had a mild case, or were infected with matter taken from their spots.

The death rate was one-tenth that of the normal disease. But this practice could cause severe disease, even new epidemics.

An English doctor called Edward Jenner lived in the beautiful Vale of Berkeley. He was fascinated by an 'old wives' tale'.

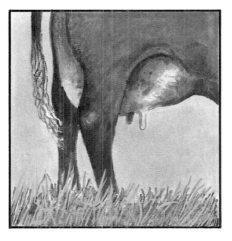

If farmworkers caught cowpox (an unsightly but harmless disease of cows' udders) they would **never** catch smallpox.

Instead of infecting people with mild cases of deadly smallpox, Jenner decided to try giving them harmless cowpox instead.

England 1796

Jenner decided to experiment on a healthy boy called James Phipps. He was the son of a farm labourer.

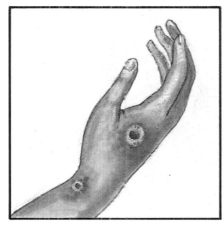

Another patient of his, a milkmaid called Sarah Nemes, had caught cowpox through a thorn scratch on her hand.

then on 14th May 1796

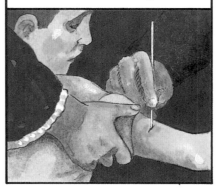

Jenner made two cuts in James's arm, each about half an inch long. In these he put fluid from the spots on Sarah's hand.

4 days later...

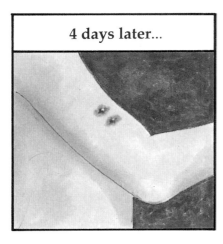

Two reddish spots appeared on James's arm. He became slightly feverish, but soon recovered.

Jenner wrote excitedly...

...'The Boy has since been inoculated for the small pox which as I ventured to predict produc'd no effect. I shall now pursue my Experiments with redoubled ardor'...

Jenner called his treatment vaccination (vacca is Latin for cow) and predicted *'the annihilation of smallpox must be the final result of this practice'*.

For about 100 years people all over the world were protected from smallpox in this way, but spreading the virus from person to person was not ideal.

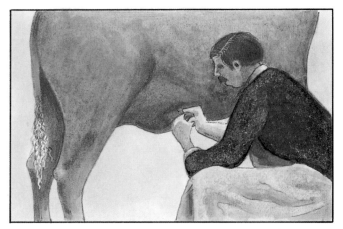

Then the vaccine was made by infecting the shaved and scarred flank of a calf with cowpox. It was usually kept in small glass bottles. In some countries the calf was taken from house to house!

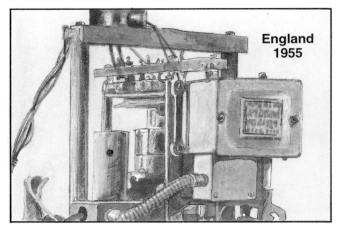

Storage of the vaccine had always been a problem. A scientist partially purified the virus and made a very stable freeze-dried vaccine.

With the modest budget of 2.5 million dollars, a World Health Assembly vowed to eradicate smallpox from the entire planet by 1976.

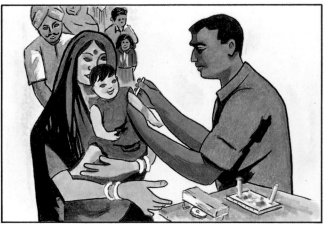

In 1967, there were probably 10 million cases a year in 44 countries. 250 million vaccines were needed each year. Many countries donated vaccines.

Smallpox was now a problem in 16 countries, mainly Asian. With 100,000 health workers, rewards for reporting cases, and armed guards to prevent sufferers infecting others, the incidence declined.

Last reported cases of smallpox:

1971   Brazil
1972   Afghanistan
1974   Pakistan
1975   Nepal
1975   India...was the world really free of this terrible disease?

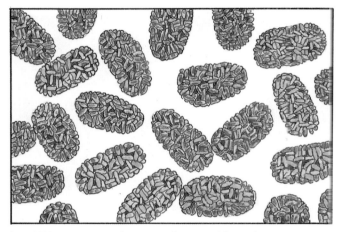

In 1978, a young photographer working above a virology lab in Birmingham, England, was accidentally infected with the virus. She subsequently died. Was she to be the last victim of this deadly virus?

In deep-frozen state and under great security, stocks of the virus are still kept in Atlanta, USA, and Moscow, Russia, while the experts debate whether to destroy them. What would you do?

The cost of eliminating smallpox was very modest compared with the money spent on weapons of war, and required co-operation between many countries who had traditionally been enemies. It was truly a world-wide campaign. It was also successful because the vaccine was effective and easily given; the virus did not seem to infect any animals; the symptoms of the disease were easy to spot; and the vaccine protected for many many years. Lastly, it was such a dreaded disease that most people wanted to be vaccinated.

Vaccines are one of the most important inventions ever made by human beings. Once scientists learned how to grow viruses and bacteria in the laboratory, they were able to make vaccines against many diseases. Some were made from microbes rendered harmless by chemicals; others were made from isolated parts of microbes or from inactivated bacterial poisons. Viruses can sometimes be weakened by growing them in certain cell cultures. Weakened viruses make good vaccines as well. You will already have been given many of the vaccines developed in the last hundred years.

More than half of all the children in the world are now given vaccines against whooping cough, diphtheria, tetanus, polio, measles and tuberculosis. Let's hope all children can be reached soon. Then, like smallpox, these diseases could be wiped off the planet. The next targets are the extinction of polio by the year 2000, and measles shortly after.

Vaccination is much easier and cheaper than many other treatments. But vaccines have not yet been made against all the most dangerous diseases, for instance AIDS. It is not always possible to safely inactivate or weaken the germ. Some germs, especially viruses, keep on changing, so vaccines will have to keep changing as well. And some infections, for instance colds, are caused by hundreds of slightly different viruses, so vaccines will be needed against each one.

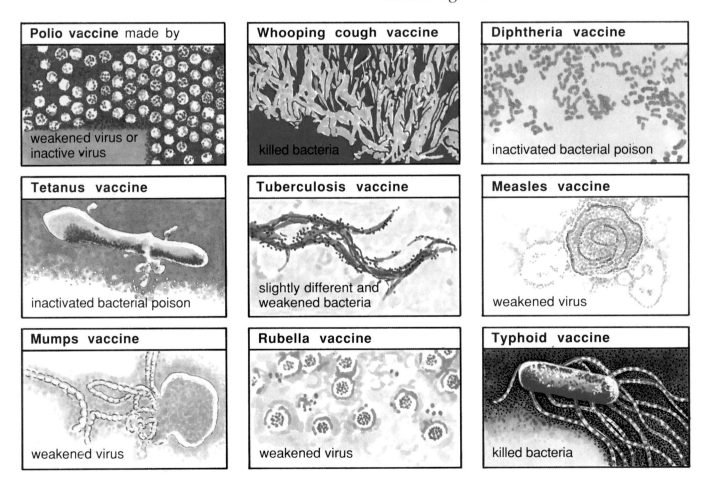

Polio vaccine made by
weakened virus or inactive virus

Whooping cough vaccine
killed bacteria

Diphtheria vaccine
inactivated bacterial poison

Tetanus vaccine
inactivated bacterial poison

Tuberculosis vaccine
slightly different and weakened bacteria

Measles vaccine
weakened virus

Mumps vaccine
weakened virus

Rubella vaccine
weakened virus

Typhoid vaccine
killed bacteria

Vaccination works because defender cells that protect you from infection can remember a germ they have met before. In this respect, the most important defender cells are lymphocytes. When a lymphocyte squad first meets the germ it is destined to fight, it begins to multiply rapidly. After a few days some lymphocytes make special protein weapons called antibodies that neutralize the germ. Other lymphocytes become killer cells, attacking the germ or cells infected by it. When the infection is defeated, a few memory lymphocytes remain on alert for that particular germ, for weeks, months, and even years. If the germ invades again, the defender cells respond faster and more effectively.

Vaccines make lymphocytes react as if they are fighting the real live germ instead of a dummy. When the real germ attacks, lymphocytes are already on alert and the germs are destroyed in double quick time, before they cause damage.

Jenner didn't know any of this when he vaccinated James Phipps, but as we understand more about our immune systems and the diseases they fight, we should be able to make better and safer vaccines. They are certainly needed. In the richer countries of the world, less than one-tenth of people now die of infections, but in poorer countries one-third to one-half die of these diseases, usually before they are 14 years old.

Vaccination, antibiotics, drugs from the chemical industry, living conditions, and resistant bacteria, all influence the ability of the human race to eradicate infections. There is one disease that illustrates this particularly well...

...its story is a powerful example of success and failure. In 1821, at the age of 25, the poet John Keats lay dying from consumption. This disease attacked rich and poor, and seemed to threaten the civilization of 19th century Europe, by killing one-quarter of its population. We now call this bacterial disease tuberculosis, or TB. Even in the 20th century, TB has killed more people than any other bacterium.

The TB bacterium has a tough, waxy outer wall. It is most commonly spread by droplets in the air. The bacteria are usually destroyed by macrophage defender cells in the lungs of healthy, well-nourished people. But the bacteria may grow inside macrophages, which eventually burst open, releasing millions more bacteria. Other defender cells try to wall off the bacteria, making a small lump. Sometimes these lumps heal, but the bacteria may multiply and break through, spreading all round the body. Most of the distressing symptoms of fever, weakness and terrible coughing are caused by the massive reaction of defender cells to the invading organisms.

The turn of the century brought the first hope in the fight against this most serious disease. Not 'wonder drugs', but simple improvements in living conditions, less overcrowding and better diets. Stronger and fitter people could fight the bacteria. Those who caught TB were often sent away to hospitals in the countryside or mountains, for rest cures.

The second weapon was BCG vaccine, made from a closely related cow bacterium, but this was not popular in some countries, particularly America.

In the 1940s, TB was still a potential death sentence. Penicillin was no use. There was great excitement about a new antibiotic, streptomycin, isolated in 1944 from bacteria in heavily manured soil. It was difficult and expensive to make, but seemed to be the 'wonder drug' the world had waited for. There were miracle cures of dying patients, but TB bacteria had an extraordinary ability to become resistant. About the same time, a young Swedish chemist had an inspiration that a chemical change to one of the world's commonest drugs, aspirin, would provide a cure. Tests quickly showed he was right, but once again the bacteria developed resistance. From the German chemical industry came a third 'wonder drug', a cheap and easy-to-make chemical, isoniazid. Some bacteria became resistant to this as well!

The battle was finally won in developed countries by isolating and treating infected people with a combination of the three drugs. Milk was pasteurized to stop infection from cows; mass X-ray screening of whole populations detected undiscovered infections; and, in Europe, vaccination of young people gave them important resistance. One by one the large TB hospitals became empty.

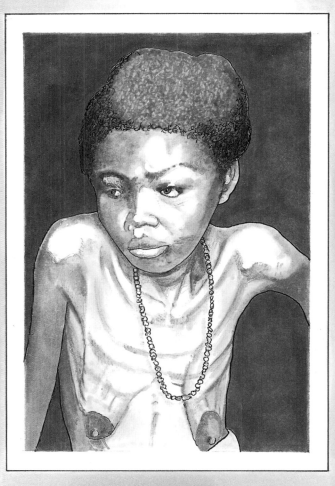

There was cautious optimism. However, by 1990, a worrying increase in cases was recorded, not just in the poorer countries, but also in places like New York. Many of these cases were connected with the newly discovered HIV virus that causes AIDS. HIV infection seemed to activate TB and many cases were resistant to most drugs. By the year 2000 more than half the world's population will live in cities. The resulting over-crowding and poverty will make such cities a breeding ground for TB. The disease is still a major threat as 1.7 billion people still carry the bacterium. TB is a global time bomb for the 21st century...

Not all your enemies are as small as viruses, or bacteria. Creatures that creep, crawl, wriggle, and even fly, can make humans ill. Tapeworms, flatworms and roundworms, each made of several million cells and visible to the naked eye, are parasites that can spend all or part of their lives living inside human beings. You may be relieved to know that there are drugs that attack and destroy these unwelcome visitors. Some give the worms uncontrollable muscular spasms and make them so weak that defender cells can destroy them. Other drugs paralyse the worms, or stop them making energy from the carbohydrates their human host has eaten. A few arthropods, joint-legged animals without backbones, are also human enemies.

Particular villains are some spiders, mites, lice, ticks, centipedes, millipedes, fleas and flies. Insect bites cause painful swelling and itching. Head lice are common human parasites that cause itching on the back of the head. They are treated with insecticides, not unlike those you might use to spray your garden to get rid of greenfly! Arthropods usually make humans ill because they carry bacteria, viruses and other microbes from person to person. They are like 'incubators', allowing the microbes to multiply in their bodies before spreading them to more human victims. Because they carry diseases these arthropods are called vectors. A bite from a vector is like an injection with a contaminated syringe! One of the very worst diseases transmitted by arthropods is malaria.

Malaria is caused by a microscopic germ called *Plasmodium*, which is injected into humans by the female *Anopheles* mosquito. Malaria is one of the world's biggest health problems, killing about two million people each year, most of them children. At any one time, there are more people ill with malaria than any other infection. *Plasmodia* have complicated life cycles as parasites of both mosquitoes and humans. In humans, acute attacks of violent fever occur each time the *Plasmodia* burst out of red blood cells. People with chronic malaria may become weak and anaemic. In the most serious type of malaria, the tiny blood vessels of the lungs and brain become badly damaged. This causes coma and one-fifth of the patients do not recover.

In the early 17th century, Jesuit missionaries learned one of the most important secrets of the Inca herbalists, the power of the Peruvian Fever Tree. Long before the cause of malaria was understood, people all over the world began to use its bark, now known to contain the chemicals quinine and cinchonine. It was not long before demand outstripped supply. In 1850, for instance, nine tons of bark were carried across the High Andes for export to the British people living in India. In this century, over half a million chemicals have been tested against malaria, but only 20 or so have proved useful. *Plasmodium* is quite different from human and bacterial cells. Drugs exploit this, often stopping the parasite making its DNA properly. The drugs can make people better and prevent infection, but, unfortunately, *Plasmodia* often become resistant to the drugs.

In 1955, the Eighth World Health Assembly voted to eradicate malaria. By 1970 they had made great progress in Europe, North and South America.

The most important target was the small, but deadly, mosquito that only flies at night. Vast areas were sprayed with insecticides, especially houses where the mosquitoes rested after feeding. Mosquito habitats were destroyed. Drugs were used carefully and effectively. People were encouraged to sleep under nets treated with insecticide. Wars, inflation, political instability and famine prevented further progress, and the mosquitoes became resistant to the insecticides. Today, malaria affects more people than in the 1960s, but it is now mainly restricted to the tropics. For instance, at least 75% of children in West Cameroon have *Plasmodia* growing in their blood cells.

For adults who live in malaria zones, outbreaks of the disease may be no worse than influenza. Children are more vulnerable, and so are tourists, travelling from malaria-free areas. If you travel to the tropics, you must take anti-malarial drugs before, during and after, your visit. And, as the parasites may be resistant to the drugs, always keep well covered at night, sleep in insect-screened houses under insecticide-coated nets, and use insect repellent. Malaria may be conquered with better living conditions and political stability of the worst affected countries. But vast areas still need to be sprayed to kill the mosquitoes. The power of molecular biology may, in time, give us vaccines. However, with such a complicated life cycle, and so many different forms of the parasite, it is unlikely that one single injection will ever protect for life. The fight against malaria is far from over.

Antibiotics and vaccination have won many battles against infection in the 20th century. But the richer nations have become complacent. Antibiotic-resistant 'superbugs', over-population, and poverty could still destroy our fragile supremacy over so many deadly microbes.

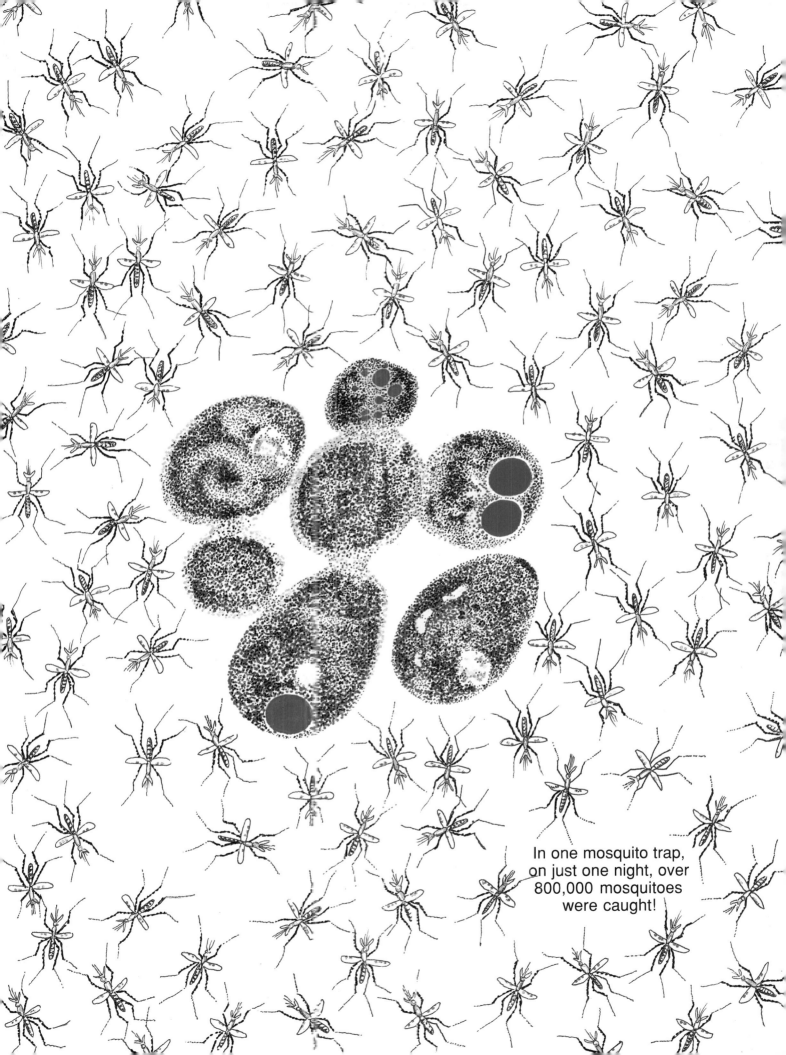

In one mosquito trap,
on just one night, over
800,000 mosquitoes
were caught!

Rebel Cells

Humans have another enemy that can make them ill. This is neither a microbe nor a bug, but a human cell that gets out of control. The disease it causes is called **cancer**.

Cancer happens because the DNA genetic information of a cell is somehow changed. It may be damaged by chemicals, high-energy rays from radioactivity or ultra-violet radiation from the sun. It may be altered if mistakes are made when the DNA is copied inside the cell.

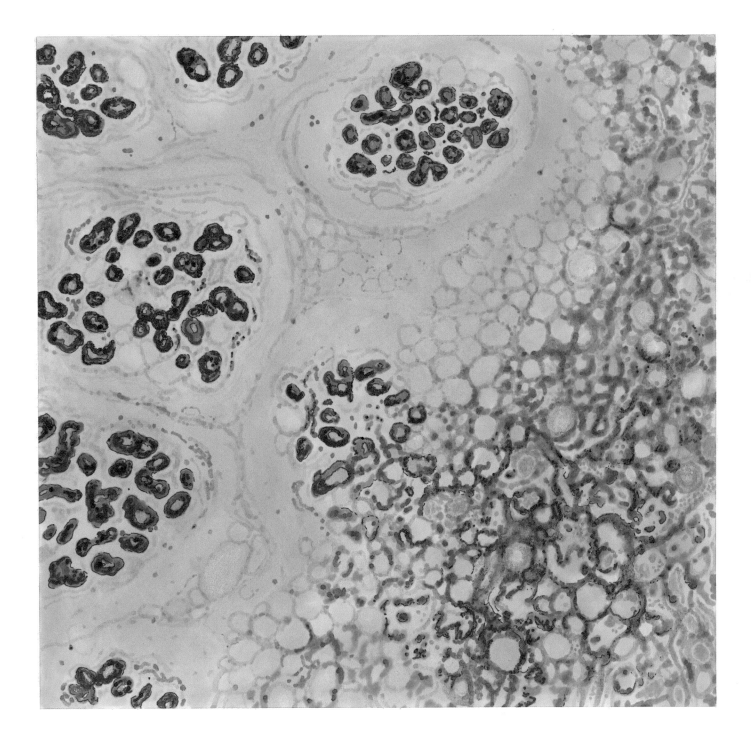

In fact, these mistakes are quite common and rarely cause trouble. Most cells with altered DNA behave quite normally and are often able to repair the damage. If the damage is really serious, the cell will self destruct. However, if the DNA that controls cell growth and multiplication is affected, that cell could become a cancer cell. Even so, a cell with damaged DNA will cause no trouble for many years. It usually needs damage to more than one part of DNA to make a cancer cell.

Eventually the altered cancer cell stops doing its particular job and begins to multiply uncontrollably, making many more cancer cells. These cancer cells don't respond to the signals that control the way normal cells behave.

The cancer cells usually grow into a lump made of many millions of cells. This is called a tumour and is probably found where the first altered cell started to grow. A tumour can often grow to be several centimetres wide before it is detected. Even then it may not cause much harm. The patient can often be cured if the tumour is removed by a surgeon.

The real problem is that cancer cells don't stay in one place. The rebel cells begin to spread away from the original tumour, using blood and lymph vessels. They invade other parts of the body, especially the liver, lungs and bone marrow, and make more tumours there. These are called metastases.

There are many types of cancer because there are many different types of cell within the human body. Breast cancer, for instance is caused by rebel cells from the breast, lung cancer, by rebel lung cells and leukaemia by cancerous white blood cells in the bone marrow.

Cancer is most common in older people and rare in children. This is because a cancer cell usually takes many, many years to get out of control. Only 1 child in every 650 will get cancer between birth and 16 years. There are some simple things you can do that will reduce your chances of suffering from cancer as an adult.

Don't smoke

Cigarette smoke is full of chemicals that damage DNA. Smoking doubles your chances of getting any cancer, and you are eleven times more likely to get lung cancer than someone who does not smoke. Smoking also increases the risk of heart attacks by damaging cells that line blood vessels.

Protect your skin from the sun

Rays from the sun can damage the DNA of your skin and eye cells. Wear sunglasses and a hat in the sun. Protect your skin cells with cream that filters out the harmful rays from the sun. Paler skins need stronger sun creams.

Eat plenty of fresh fruit and vegetables

Citrus fruit, in particular, contains substances that protect your cells from DNA damage. During their normal activities, cells make some waste products that may damage DNA. Chemicals in citrus fruit, particularly vitamin A, can mop up these damaging waste products.

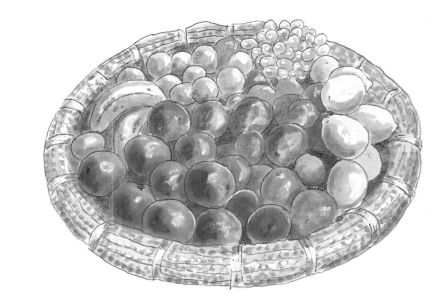

It is difficult to find cures for cancer because the cause of the trouble is human cells. There are so few differences between cancer cells and their normal counterparts that can be exploited in treatment. From ancient times doctors and scientists have wrestled with this problem.

All this may sound rather frightening, but over half the people who get cancer can be cured, especially if the cancer is found before it has spread around the body. Scientists and doctors understand much more about cancer than they did even five years ago. The more they understand about cancer cells, the easier it will be to cure the disease.

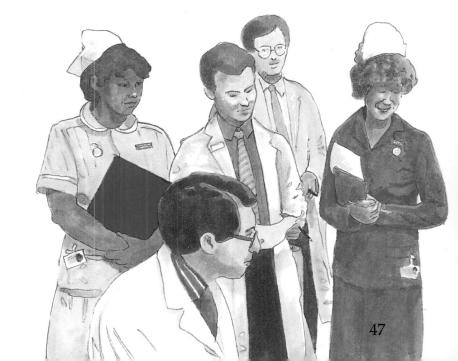

One of the treatments we have today exploits the fact that cancer cells multiply more rapidly than normal cells. The origin of this treatment goes back to soldiers on the battlefields of the First World War, who were dying from a deadly silent weapon of chemical warfare, poisonous nitrogen mustard gas.

Doctors examining the bodies of the dead soldiers noticed a lot of damage to cells in their bone marrow. They also had very few white blood cells.

No one thought much about this until after the Second World War when scientists working on chemical weapons wondered if the nitrogen mustard gas might have killed the bone marrow cells because these cells multiplied rapidly. If this was true, they reasoned, similar chemicals might kill cancer cells. They made some trial drugs and gave them to mice who had the white blood cell cancer, leukaemia. It was extremely successful, the mice were cured.

This was the beginning of cancer chemotherapy. Scientists and doctors screened tens of thousands of natural and laboratory-made chemicals looking for anything that would kill multiplying cells. But there were drawbacks to this approach.

The drugs that were discovered usually killed all rapidly multiplying cells in the body. This meant that they killed some of the normal cells that lined the stomach and intestines, cells that made hair, and cells in the bone marrow.

Another problem was that only about 5% of the cells in a tumour are multiplying at any one time. Cells that are not multiplying cannot be killed by this treatment.

As well as all of these drawbacks, it was soon clear that chemotherapy would only work against some types of cancer.

In spite of these problems there have been some important successes, particularly in treating children who have cancer. Each year, about 1300 children in the UK will develop cancer. Before 1970, all but 100 of those children would have died. Now over half are cured, and this is mainly because of cancer chemotherapy.

One of the most remarkable stories is the cure of leukaemia, bone marrow cell cancer, in children. The cure consists of a powerful cocktail of drugs made from extracts of a common garden plant called the periwinkle, antibiotics and hormones...

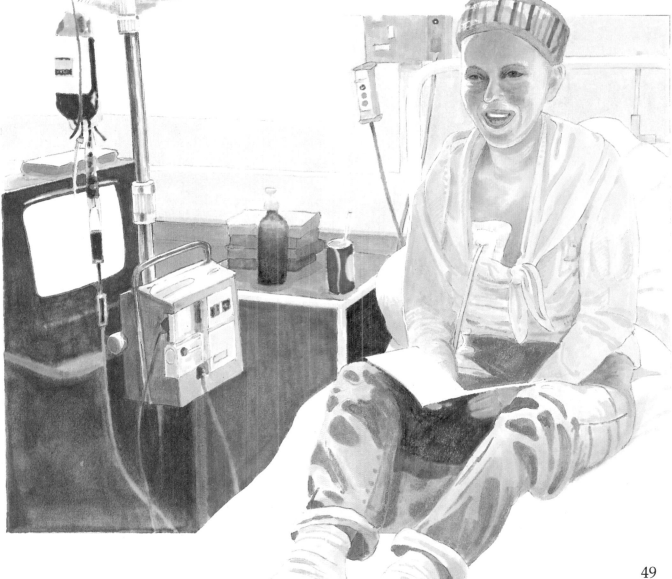

The periwinkle is an evergreen shrub that grows throughout the world. The delicate star-shaped flowers are pink or white.

For centuries, the native Brazilians used an infusion of periwinkle leaves to clean and heal wounds, to stop bleeding, and as a mouthwash for toothache.

In the Philippines and the West Indies, native doctors used extracts of periwinkle to treat people with a disease called diabetes.

In the first half of the 20th century, there was no cure for diabetes, so scientists decided to investigate this folk remedy. They extracted many different chemicals from the plant and purified them.

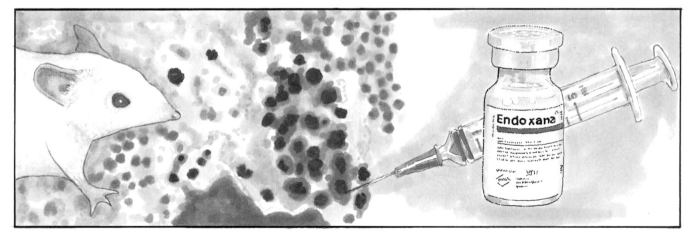

By 1957 the extracts were ready to be tested on mice to see if they would cure diabetes. None of them worked. Moreover, some of the extracts were very poisonous to bone marrow cells.

Might these be used to treat leukaemia instead of diabetes? The extracts cured leukaemic mice. One was made into a drug called vincristine. It was soon the most successful drug for children with leukaemia.

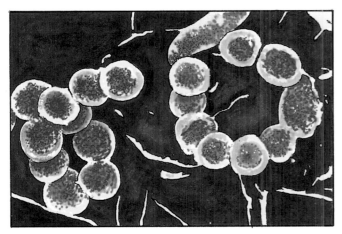

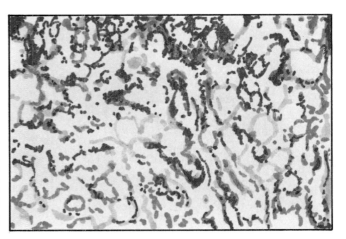

Other drugs had been discovered that would kill human cells. Some of these were antibiotics that were originally developed for infections. One of the best came from bacteria found in Sardinian sewage!

Between 1950 and 1970, scientists also found that some hormone-like drugs made in the laboratory, were particularly good at stopping cancer cells dividing. But in spite of all these new drugs, few patients were cured.

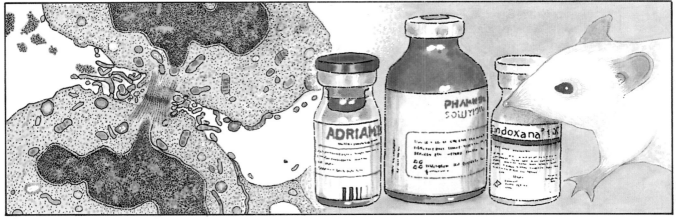

Each of the drugs would kill about 99% of the cancer cells, but the cancer cells remaining could grow again and, like bacteria, become resistant to the drug.

The solution was to combine the drugs. This was difficult and dangerous. Experiments in mice proved that combinations killed 99.9999% of cancer cells. Would it work in humans?

Doctors and scientists carefully compared different combinations, first in laboratory experiments and then in patients. It took many years and many thousands of patients before they had the answer.

Combination chemotherapy is now a useful treatment for some types of cancer, especially leukaemia in children.

Leukaemia is a cancer of white blood cells that grow out of control in the bone marrow. It is the commonest cancer in children. As there are billions of cells in the bone marrow, a child can feel well until it has as many as one million, million leukaemia cells. To destroy so many cells will take six or more of the most powerful drugs that the doctors can use, including vincristine...

A doctor may suspect leukaemia if a child is pale, tired, has unexplained fevers, or bleeding and bruising.

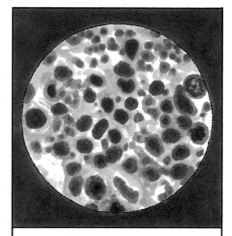

The correct diagnosis is vital. Samples of blood and bone marrow are examined by the hospital specialist.

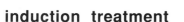
induction treatment

Speed is of the essence. The child must quickly be admitted to a hospital where children with cancer are treated.

soon afterwards

Drugs that kill cells are given in a combination and sequence that has been worked out by careful experiments.

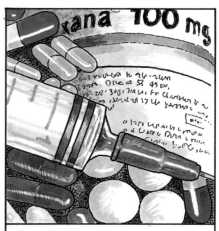

The aim is to attack the cancer cells in several different ways simultaneously.

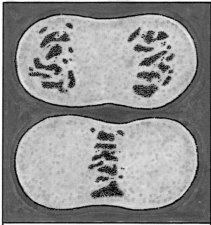

First the child is given vincristine, which stops the chromosomes being pulled apart when a cell divides.

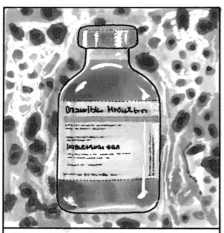

The child is also given a drug that stops cancer cells making all their proteins,

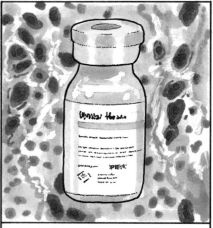

and a hormone that stops the cancer cells growing.

one month later

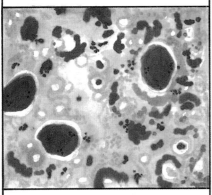

99.99% of the leukaemia cells have been killed. Most normal cells are not affected. The child begins to feel better.

consolidation

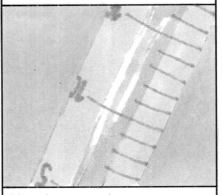

New drugs are given to mop up the remaining cancer cells. Usually one drug makes their DNA strands stick together.

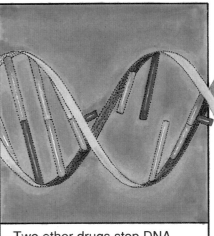

Two other drugs stop DNA being made properly in multiplying cells.

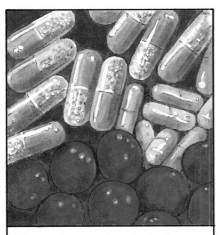

These drugs are quite toxic to normal cells, e.g. those lining the intestine, but there are drugs that help sickness.

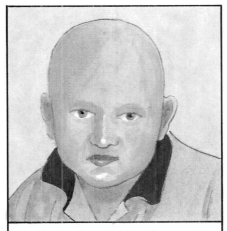

The hair cells are also damaged, but the hair will grow back very quickly, once the treatment stops.

three months later

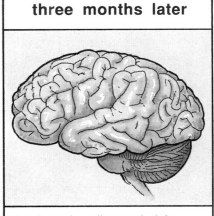

Leukaemia cells may lurk in the brain, so some treatment is given into the spinal cord.

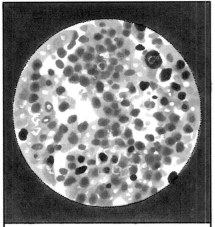

Blood and bone marrow tests show the cancer has been destroyed, but to make sure, drugs are given once again.

The treatment is almost over, the patient back at school. Chemotherapy will be given once a month for two years.

As recently as 1965, only 1% of children with leukaemia survived. In the last 30 years, experiments with chemotherapy in the laboratory and trials in hospitals specializing in childhood cancer have changed this. Now over 80% of the children who have leukaemia are cured of their illness. Some have even become doctors themselves!

Cancer chemotherapy has certainly cured some patients and given many others extra years of life, but new cancer treatments are needed, particularly for the cancers that adults get. While chance and luck will probably play a part, new cures for cancer are most likely to be discovered as scientists and doctors understand the differences between normal and cancer cells.

One exciting prospect is vaccination. The idea is to encourage human defender cells, especially lymphocytes, to recognize cancer cells as unwanted intruders and destroy them. Cells continually communicate with lymphocytes by 'displaying' some of their inside proteins on their outsides. If the proteins are normal, the lymphocytes leave that cell alone.

If they are different, for example, when a cell is infected with a virus, the lymphocytes will move in and kill. Cancer cells will have altered proteins because their DNA has been damaged, and these may be displayed on the surface of the cancer cell.

Now some patients are being vaccinated with a harmless preparation from their own tumour cells. It is not known whether this will encourage lymphocytes to destroy tumour cells in other parts of the body, but it is possible that we may, in the future, be able to vaccinate against some types of cancer as well as measles and mumps.

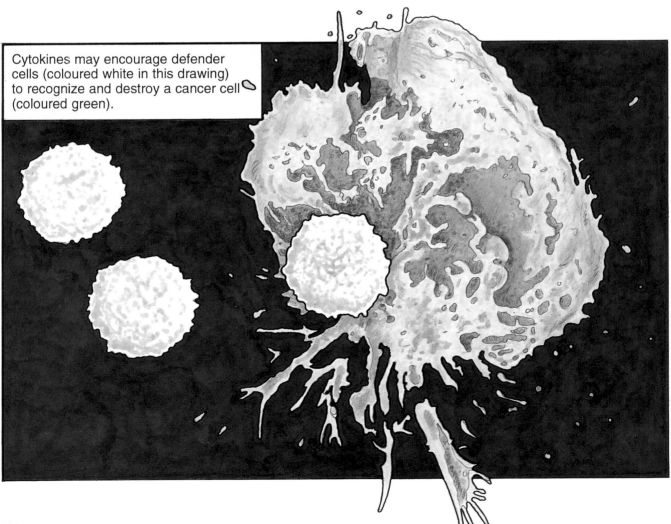

Cytokines may encourage defender cells (coloured white in this drawing) to recognize and destroy a cancer cell (coloured green).

Scientists now know a lot about the genes that are damaged in cancer cells. These genes usually control the way cells divide, die, or repair their DNA. In the future scientists may be able to develop drugs that replace or repair these mutant genes, or replace the proteins that they make.

Understanding the messenger molecules that allow one cell to tell another how to behave, may also lead to useful treatments. Some cancers, for instance, have become dependent on the body's normal hormones to keep on growing. 'Dummy' hormone drugs are made that block the normal hormone signalling the cancer cells. In breast cancer, a drug called tamoxifen, competes with the natural hormone oestrogen.

A recently discovered group of messenger molecules, cytokines, may also be useful at treating cancers. Some cytokines may be able to control the way cancer cells behave, and others may help the body's defender cells to recognize and destroy the cancer cells. One cytokine, interferon, also used in some virus infections, is a useful treatment in some of the rarer cancers.

Cancer cells mainly cause trouble when they spread around the body. Scientists now know that tissue-dissolving enzymes help this spread. There are new drugs that stop these enzymes working. A cancer that cannot spread may well be harmless.

Fifty years ago, the prospect of understanding, curing and preventing cancer seemed remote. World-wide research by dedicated scientists and doctors has changed all that.

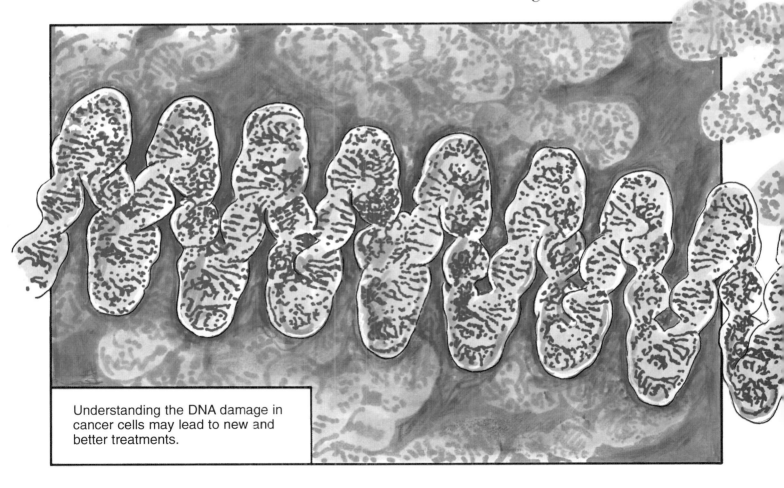

Understanding the DNA damage in cancer cells may lead to new and better treatments.

A Dark Side
of the Immune

The air you breathe, the food you eat
and the objects you touch, all carry
microbes and bugs that can invade your
body and cause you harm. Fortunately,
your immune system usually recognizes
the danger. Every day, its powerful,
versatile cells and chemical weapons
defend you against viruses, bacteria and
other microscopic enemies. Most of the
time this works with incredible
efficiency. Even if you succumb to
an infection, the full power of your
immune system will soon be directed
at the offending invaders and you
recover quickly.

But there is a problem in having such
a powerful defence system. Sometimes
it is just too efficient and mounts a huge
attack on substances that are of no
danger to your body. Defender cells turn
into destroyer cells that can damage
your body and may even, very rarely,
kill you. This is called allergy and the
substances that cause it are called
allergens.

The sneezing, itching and watery eyes
of hay fever; the breathing problems and
wheezing of asthma; the itching and
painful skin of eczema; swelling of the
face and sickness of food intolerance;
and the rare, but sometimes fatal,
collapse after a bee or wasp sting, are all
allergies. Allergy is a Dark Side of the
Immune!

These are some of the most troublesome
allergens.

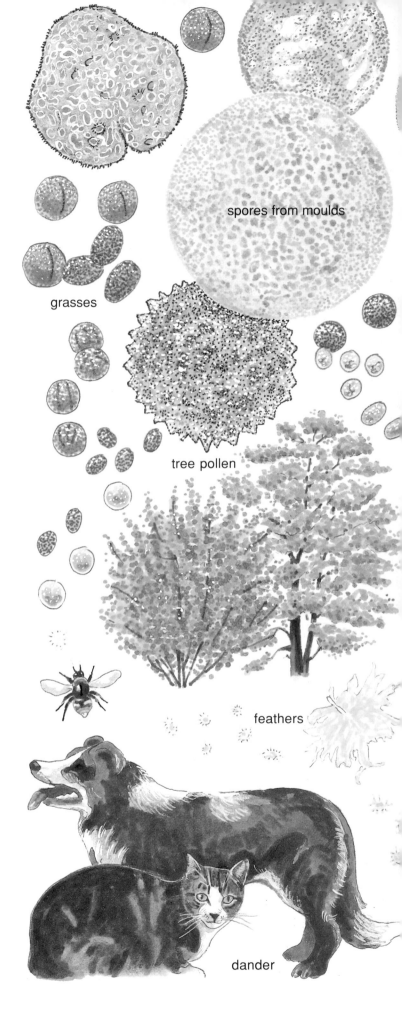

spores from moulds

grasses

tree pollen

feathers

dander

peanuts

urine and saliva from animals

cows' milk

seafood

bee venom

diesel exhaust particles

dust mite faeces

wasp stings

57

The major villains in allergy are three of the rarer types of immune cell, and an equally rare type of antibody. When allergy occurs, they work together to cause havoc in your body.

Mast cells are stationed in the moist surfaces of your body, like your nose, mouth and lungs, where your insides meet the outside. There are 10,000 mast cells in each cubic millimetre of skin, for instance. They are also found deeper inside the body, especially near blood vessels. Each mast cell carries about 1000 granules, full of potent chemicals that can cause havoc if released.

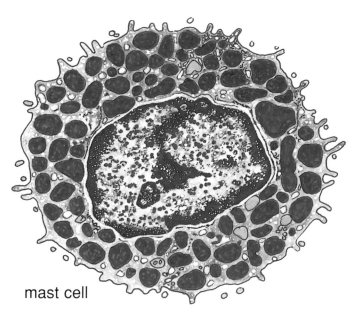

mast cell

Basophils are similar to defender cells in the blood called neutrophils, but they are much rarer. They circulate in the blood, on the watch for bacteria, viruses and other enemies which they gobble up and destroy. But unlike their neutrophil cousins, basophils also carry granules that are full of cell-destroying chemicals.

Eosinophils are also rare relatives of neutrophils. When the immune system is functioning normally, their vital job is to destroy the larvae of parasitic worms that invade tissues. In allergy, their powerful killing weapons damage human cells.

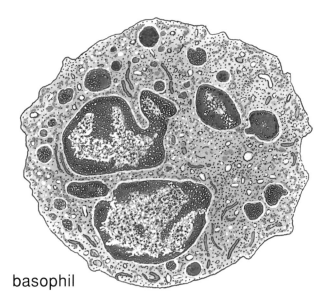

basophil

Antibodies are designer weapons made by lymphocytes. The immune system can make millions of slightly different antibody molecules depending on the enemy they are fighting. All of these antibodies are made to an overall pattern which scientists call immuno-globulin. The one that causes the trouble in allergy is called immunoglobulin E, IgE for short.

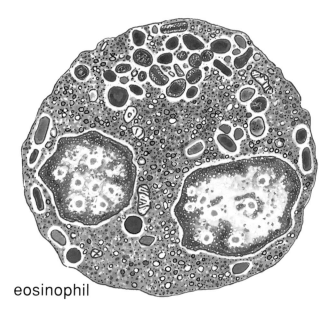

eosinophil

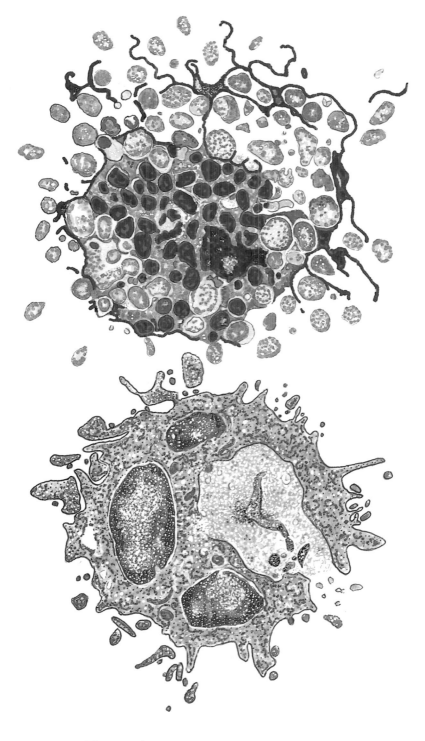

The real trouble with allergy is that IgE sticks very tightly to mast cells and basophils. It can stay there for many weeks until the cells meet with the allergen the IgE was designed to fight. When this happens, the cells explode, releasing powerful cell-destroying chemicals.

Hay fever is a good example of what happens next...

Hay fever affects about 15% of people at some time in their lives. It happens when allergens, especially pollen from trees and flowers, are breathed through the nose. Some of the proteins in these pollen grains dissolve in the watery mucus that protects the nose cells. Most of the time, the immune cells up the nose take no notice of such proteins, but occasionally they react as if the pollen grains were invading germs. The lymphocytes then make IgE antibodies to the pollen protein.

The first time the immune cells react to the pollen allergen, there is no sign of trouble. But the IgE they have made sticks tightly to the mast cells hidden among the lining cells of the nose, and the basophils that lurk in the watery mucus. The next time the same pollen is breathed in, the mast cells and basophils explode!

The pollen protein fits tightly to the matching IgE molecules on the mast cell or basophil. This unlocks the power of mast cells and makes them release powerful chemicals. Very rapidly, these chemicals widen blood vessels, cause the surrounding tissue to swell with fluid and irritate the nerve endings. This causes sneezing, runny nose, itchy and watery eyes, and will happen whenever the particular allergen is in the air. And it does not take very much allergen to start the whole thing off. We only breathe about one-millionth of a gram of pollen each year, less in weight than a speck of dust!

One of the most powerful chemicals released by mast cells and basophils is called histamine. This binds to special receptor molecules on muscle, blood vessels and nerve cells nearby to cause the symptoms of hay fever. The best drugs for treating hay fever stop the powerful histamine working. These drugs are called anti-histamines and are a similar shape to histamine. They stick tightly to histamine receptors, and stop histamine itself working. They can be taken as a pill and their effects are long lasting. Anti-histamines are very effective and useful drugs, but they can make people sleepy. This is because they also block some receptors on nerve cells in the brain, where histamine is one of the chemicals used by nerve cells to transmit their messages.

About 5% of people in the Western world are asthma sufferers. In asthma, the tiny airways of the lungs become blocked, usually for a short while. The disease is especially common in young children. This is because they breathe in more air relative to their body weight, than adults, they exercise more, and have narrower airways. 13% of children in the USA and 11% of children in the UK have asthma. In these countries, it is the most common chronic disease of children. In other parts of the world, like India, Japan and rural Gambia, very few children suffer from asthma. The reason for these differences is that asthma is an allergic disease of the lungs. The risk of children having asthma is affected by their race, and the climate, altitude and conditions that they live in.

Pollen and spores from fungi can trigger an attack, but it can also be caused by many substances in the home, especially the faeces of a minute creature called the house dust mite. Feathers, animal fur and some foods like peanuts can cause asthma. Cigarette smoke, air pollution, cold air, exercise and lung infections can make asthma worse. In asthma, IgE antibodies are stuck to mast cells in the lungs. When the allergen is breathed in, it binds to its IgE and causes the mast cells to release histamine and other powerful substances. These immediately cause spasms in the muscles of the tiny airways, so that it is difficult to breathe out. The sufferer usually gasps, wheezes and coughs as air is forced through the narrowed tubes.

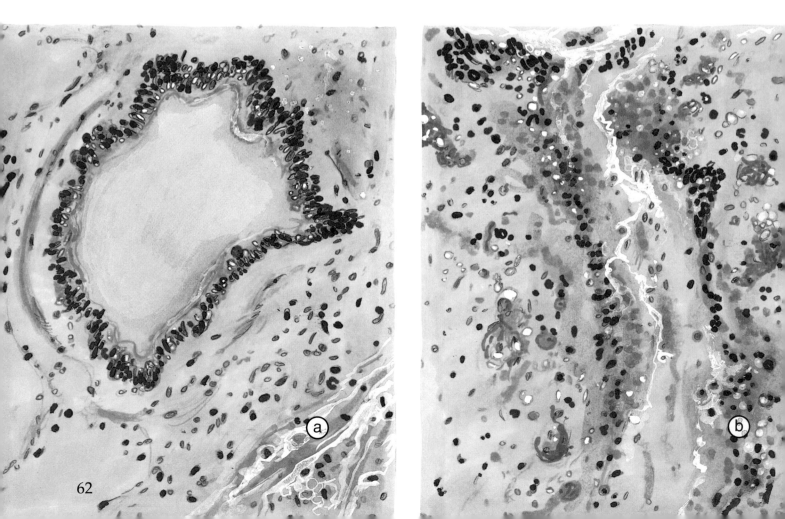

Other chemicals released by the mast cells cause defender cells to flood into the lungs, causing inflammation. Eosinophils, in particular, destroy some of the lining cells of the lung. Very soon the lungs of asthma sufferers become ultra-sensitive to all kinds of changes in the air they breathe.

Drugs to treat asthma are among the most widely prescribed medicines for children and teenagers. They either work to stop the immediate effects of the damaging mast cell chemicals, or the later effects of the inflammation. Many children need to take both types of drug to control their symptoms.

Rapid relief is given by drugs that open the tightened airways. These work by mimicking a natural hormone that the body uses to relax muscles like those in the tiny air tubes of the lungs. The hormone is called adrenaline, and when it is released normally, it makes the heart beat faster and the muscles of the lung relax so that more air can enter. For over 80 years, the hormone adrenaline was used to treat asthma. More recently, scientists have been able to design drugs that act on the lung muscles, with less effect on the heart.

The most usual drug of this type is called salbutamol. The best way to get this drug quickly to the lungs is to breathe it in using a simple inhaler or a more complicated nebulizer. One 'puff' of the inhaler gives a measured dose of drug that is breathed into the lungs. If this does not work efficiently, then a nebulizer gives a powerful aerosol spray of the drug to the lungs.

(a) a slice through a small air tube in the lung

(b) a slice through a narrowed air tube in the lungs of an asthma patient

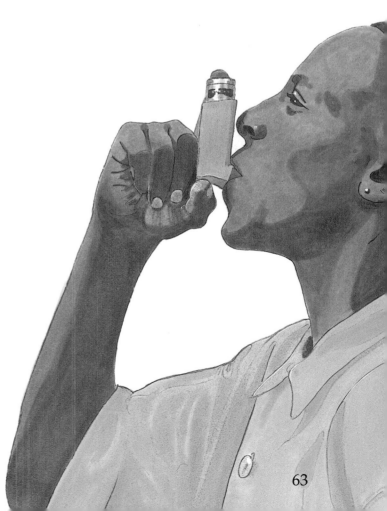

Another good way to treat asthma would be to prevent the mast cell from releasing its dangerous chemicals. This may be how drugs like cromolyn (Intal) work. This drug was developed by Dr Roger Altounyan, a scientist and asthma sufferer. He was convinced that a plant, called khellin, which was used in ancient Egypt to relax muscles, could help his asthma. He made many extracts, and deliberately inhaled them to see if they would prevent his attacks. Several years and 2000 asthma attacks later, he isolated cromolyn. But the new drug was very insoluble. Inspired by the planes he had piloted in the Second World War, Altounyan developed a small propeller-driven device to get the drug to the lungs. This drug prevents asthma attacks rather than relieving their symptoms.

Other asthma drugs stop defender cells getting into the lungs and causing damage. They resemble some hormones, called glucocorticoids, naturally produced in the body during inflammation. One of the jobs they do in the body is damp down the immune system.

Scientists have made chemicals that act like these hormones. They stop immune cells making many of their powerful chemicals, but, most importantly, stop them invading the lungs. These drugs are very powerful, and they can have quite serious side-effects, particularly on the immune system and growth in children. But, as with the other asthma drugs, they can be delivered straight to the lungs using an inhaler, and then they are much safer.

Asthma and hay fever are two of the most common allergies that can make children ill. The skin disease eczema also causes suffering in children. This may be caused by allergy to house dust mites, cows' milk, eggs, fish, nuts, or pollen, but often the cause is not properly understood.

There are other ways to treat allergies. The most simple is to avoid the allergen, often easy if it is a food, but difficult if it is in the air, like pollen or dust mite faeces.

Another way is to alter the way the body reacts to the allergen. This is called immunotherapy and is done by injecting the patient with very tiny quantities of the allergen over several weeks or months. In this way, immune cells can be made to 'tolerate' the allergen and no longer mount such a massive attack when they meet it naturally. Immunotherapy certainly works, but can only be used in very serious cases. This is because it may provoke such a powerful allergic response that the patient's life can be endangered.

The good news is that many children with allergies, particularly asthma and eczema, get better as their immune system grows older, and (hopefully) wiser!

What a Relief

Have you ever tried to hammer a nail and hit your thumb instead?

You feel a sharp, intense pain and jump around in agony. This is because nerve cells called nocioceptors respond very quickly to chemicals released by your damaged cells. They send signals along two types of nerve fibre. One type carries the pain signal very rapidly, at a speed of 320 kilometres per hour. The message is returned from the spinal cord immediately. The return signal makes your muscles move your thumb almost instantly away from the source of pain.

After the initial agony, your thumb throbs with dull pain. This is caused by signals carried along slower nerve fibres. These signals make your brain tell you that your thumb hurts.

For a while it will be painful if you try to use your thumb. The nerve endings in the bruised tissue send pain signals at the slightest changes in pressure. This stops you damaging the thumb any more and helps healing.

It may not seem obvious when you have trapped your finger in a door or bitten your tongue, but pain helps you protect your body by making it very unpleasant if you damage it. Nerves in the skin respond to heat, sharp objects, pressure and itching. Being tickled can feel excruciating because nerve endings are being mildly but constantly stimulated.

66

Many other parts of your body have nocioceptors which send powerful pain messages to the brain. Ask anybody who has had appendicitis! Unfortunately, you can't move your body away from a source of pain when the discomfort is caused by illness.

An ache or pain is a way of sending a message that something is wrong. Doctors may need to understand that message. That is why they gently (or not so gently!) prod your body to find the site of the pain. This may be uncomfortable, but it is often the first step in making you better. Quite often you feel pain some distance from where the trouble is. That is because different parts of your insides may share the same nerves. For instance, problems under the diaphragm can give you a pain in the tip of your shoulder.

The sensation of pain can be divided into four different elements:

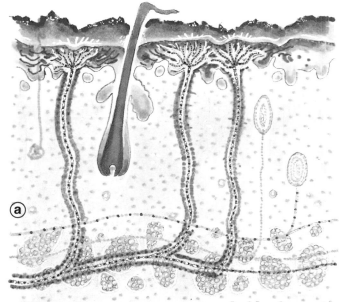

(a)

there are messages from the nerves near to the damaged cells

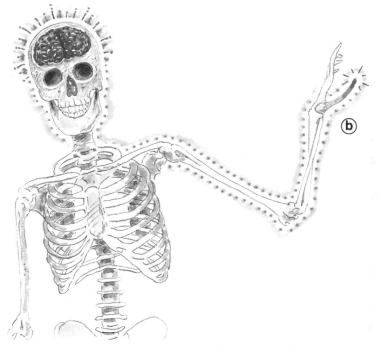

(b)

these messages are relayed to the spinal cord and on to the brain

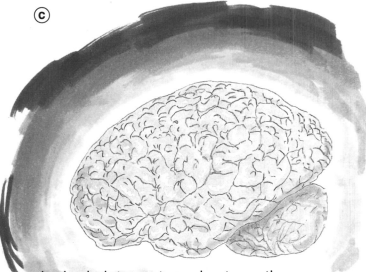

(c)

the brain interprets and acts on the messages received

and finally there is the way you feel about the pain. How you react to pain depends on many things: previous experiences, your mood, and how anxious you are. From ancient times, one of the most important jobs in caring for the sick has been to control and relieve pain. Drugs have been discovered that can work at all these different levels. One of the most effective painkillers has been used for several thousand years...

You might think that leaves and hot water make tea, but for thousands of years mixing flowers, leaves or bark with water was a common way to make drugs, particularly for pain. In 1500 BC an Egyptian papyrus prescribed dried myrtle leaves for abdominal pain. 2000 years ago the famous Greek doctor Hippocrates made juice from willow bark to relieve pain and fever.

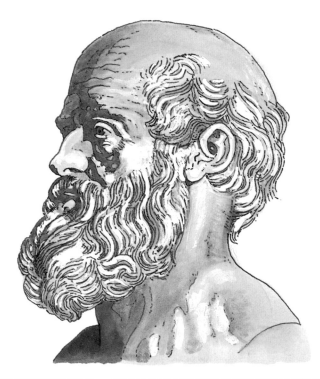

Europe The Middle Ages

Herbalists made a bitter concoction of willow bark to treat pain, but most people used willow to make baskets instead!

Oxfordshire England 1757

The Rev. Edward Stone, a country vicar, tasted some willow bark one day, and found it had a bitter and unusual taste.

Herbal cures were thought to grow near the cause of disease. Willow grew in moist places where fevers were common.

Stone began to experiment. He collected one pound of bark, dried it by a baker's oven for three months and crushed it to a light brown powder. This he mixed with water, tea or beer.

He cautiously began treating patients. Seeing some benefit and causing no harm, he increased the dose. The results were dramatic, the bark potion cured aches and fevers.

London England 1763

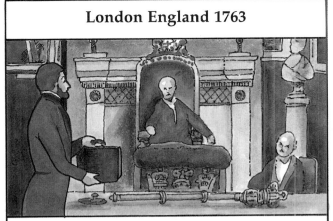

Having treated over 50 patients, Stone felt confident enough to present his results to the distinguished scientists of the Royal Society. They were suitably impressed.

40 years later...

Napoleon's continental blockade stopped supplies of fever medicines from South America. Scientists turned to willow, and tried to extract the active ingredient.

Munich Germany 1828

Professor Buchner extracted a yellow substance he called salicin from willow bark. The next year, a French chemist, Leroux, further purified salicin to a crystal form.

The Sorbonne Paris 1838

An Italian chemist, Pira, finally extracted the active ingredient. The colourless crystals did everything willow bark could. He named it salicylic acid from the Latin for willow (*Salix*).

Marburg Germany 1859

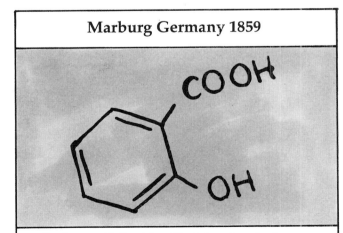

Professor Kolbe identified the chemical structure of salicylic acid. The way was clear for industrial production. His student von Heyden founded a company to do this in 1874.

Charite Hospital Berlin Germany 1876

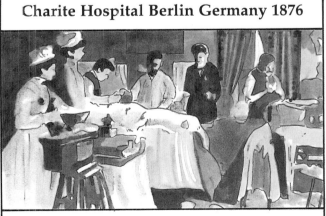

Dr Stricker experimented on patients with various diseases. He proved that salicylic acid had powerful effects against fevers, aches, pains and rheumatism.

The simple molecule was very easy and inexpensive for chemists to make in the factory. Many people used it and doctors could now prescribe accurate doses of the powder. Chemists realized that salicylic acid was also the active ingredient of pain-relieving potions from the myrtle leaves used by the Egyptians, and from the meadowsweet plant, which was used to treat pain in Medieval times.

Yuk!

But salicylic acid had a very sharp and bitter taste, and if people took it for too long, it damaged their stomachs.

Elberfeld Germany 1897

$$C_6H_4 \underset{COOH}{\overset{OCOCH_3}{<}}$$

Felix Hoffmann worked at the Bayer Dye company. His father was tortured by rheumatism, but salicylic acid made him sick. Hoffmann tried to improve the drug and finally chose a process called acetylation. This also made the drug more stable.

Bayer Dye Company Germany 1899

The new drug was useful for rheumatism, headaches, toothaches and fevers and was less damaging to the stomach. The company needed a name for this exciting drug. They may have named it after Saint Aspirinius, the patron saint of headaches! More likely they combined the German name for a chemical identical to salicylic acid found in sap of the meadowsweet plant (*Spiraea ulmeria*), and 'A' for acetyl. This gave a name that is recognized all over the world...

COOH

O-CO-CH₃

Meanwhile back in Elberfeld 1900

Aspirin was a great success, but the powder did not dissolve easily. So aspirin was made into a water-soluble tablet. It was the first major drug to be sold as a pill. Aspirin was soon used all over the world.

Journey to the Moon 1969

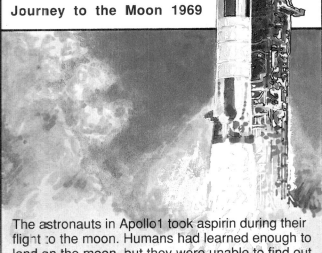

The astronauts in Apollo1 took aspirin during their flight to the moon. Humans had learned enough to land on the moon, but they were unable to find out how this common and useful drug worked.

Royal College of Surgeons England 1971

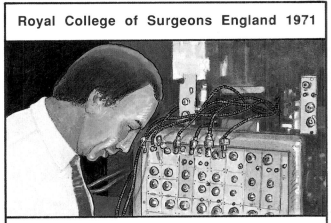

A team of scientists, led by Professor John Vane, found that aspirin bound to enzymes that normally make prostaglandins. Prostaglandins are produced naturally during infection or damage.

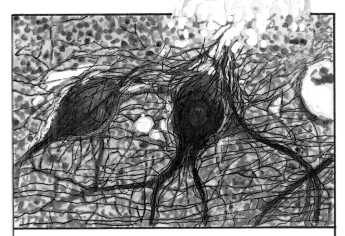

Prostaglandins, for instance, make nerve cells more sensitive to pain, particularly pain in muscles and blood vessels. Prostaglandins increase swelling when tissues are inflamed.

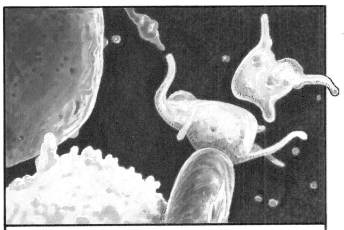

Doctors also found that aspirin can help problems with blood clotting because it stops platelets (yellow in this picture) sticking together. It may be useful in preventing heart attacks and strokes.

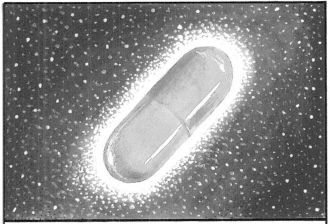

Each year about one hundred thousand million aspirin tablets are sold. Next time you see someone chewing tree bark, don't laugh, they may be about to discover the next **'Wonder Drug'**!

While aspirin is still a popular and useful drug, many people now use paracetamol (called acetominophen in the USA) if they have a fever or headache. The discovery of paracetamol did not involve a herbal remedy. Its origins are the waste products of the 19th century German dye industry, and a mistake made by a pharmacist!

In 1883 two young doctors decided to use a chemical called naphthalene to treat a dog with distemper. They sent a messenger to Strasbourg to collect the white powder. It worked extremely well, so one sent a sample to his chemist brother. The chemist was puzzled. The white powder was not naphthalene but acetanilide. Acetanilide had rather unpleasant side-effects, but it certainly reduced fevers.

Meanwhile, standing in the storage yard of the Bayer Dye Company, there was a load of barrels containing 30,000 kilograms of a waste chemical that was similar to acetanilide. Could this be converted into a useful drug? The chemists made a compound and found that it was more effective than acetanilide.

The new drug was called phenacetin and launched as an anti-fever drug in 1888, just before a major 'flu epidemic spread all over the world. Phenacetin was used for the next 90 years to treat pain and fever, but doctors began to realize that it could sometimes cause kidney damage. This was no problem because scientists had discovered that in order to work, phenacetin was converted in the body to the drug we now call paracetamol. They made paracetamol in the laboratory and found it was safer than phenacetin. Even young babies could be given it.

Like aspirin, paracetamol also stops prostaglandins being made in the body, but it does this in a slightly different way. It is important to remember however, that although aspirin and paracetamol are both very useful, they can, like all drugs, cause harm if more than the recommended dose is taken.

Next time you have a headache, you might ask yourself what is going on. It is not the nerve cells of your brain that are hurting (they have no pain receptors). A headache is probably caused because blood vessels in your brain start making prostaglandins. This makes the nerve endings in the vessels ultra-sensitive to stretching and movement. Cells in the membranes that surround the brain may also make prostaglandins; nobody really knows why.

If you swallow aspirin or paracetamol tablets, the drug passes through your stomach wall into the blood stream and gets to work within half an hour. The cells at the trouble spots cannot make prostaglandins, for a few hours at least. The nerve endings are no longer so sensitive and the pain usually goes away.

Prostaglandins are one type of molecule that make pain worse in your body. But scientists reasoned that the body may have other molecules that make nerve endings less sensitive to pain. These substances might help to control the sensation of pain. This could explain why some people suffer more pain than others, and would certainly have been of great use to our early ancestors. In times of great danger, people can seem insensitive to pain. One story comes from the famous African explorer and missionary David Livingstone on the occasion of his being mauled by a lion:

...'starting and looking half round, I saw the lion just in the act of springing upon me...Growling horribly close to my ear, he shook me as a terrier does a rat. The shock produced a stupor similar to that which seems to be felt by a mouse after the first shake of a cat. It caused a sort of dreaminess in which there was no sense of pain nor feeling of terror, though I was quite conscious of all that was happening'...

The story of your body's natural painkillers starts 6000 years ago with the milky juice of unripe poppy seed capsules. This contains one of the most powerful and useful painkillers known to man, morphine. In its impure form, morphine is called opium, and has been used since 4000 BC, not only as a painkiller but also as a drug that produces hallucinations.

74

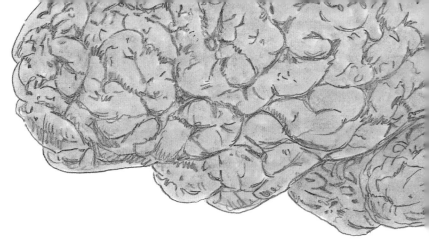

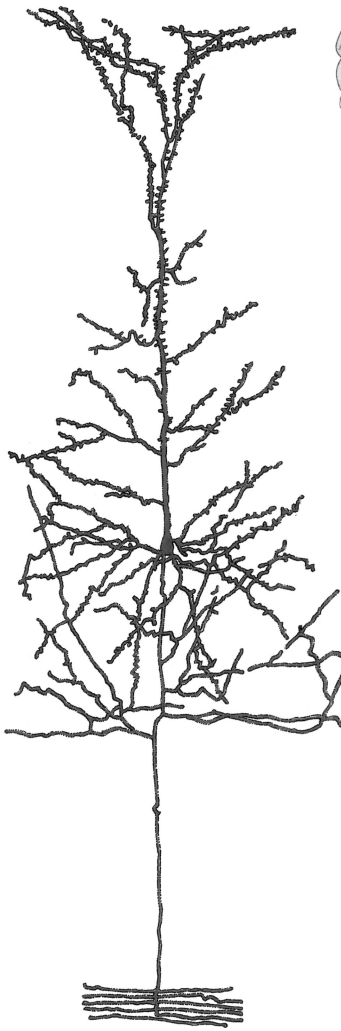

Morphine can have several different effects on the brain, but it wasn't until the 1970s that scientists found out why. There is a particular type of protein in the membrane of brain nerve cells that sticks very tightly and specifically to morphine molecules. These proteins are called opiate receptors. But why are they there? The amazing discovery was that there was a whole family of natural molecules in the brain that bound to the opiate receptors. They are now called endorphins, or natural opiates, and they are a very similar shape to morphine. They are, in fact, the body's own natural painkillers.

How do endorphins work? Scientists explain this with the 'gate control theory'. Put simply, it seems that nerve signals caused by pain have to pass through a series of 'gates' where the strength of each signal is measured. If the pain signal is strong enough it will carry on up to the next gate and eventually to the brain.

The more endorphins you make, the less pain you feel. Stress, fear, excitement and danger stimulate the production of endorphins in the brain. The more of these natural painkillers you make, the more difficult it is for signals from pain, heat and pressure nerve endings to get through the gates, and the less pain you feel.

Doctors do not use morphine for all pain because it has many serious side-effects, but there may be a way of using natural opiates. Hundreds of years ago, the Chinese discovered that stimulating some areas of the skin and underlying tissues could relieve pain and cure some illnesses. At first, they may have used sharp stones, but later they used fine bone, or needles. This is what acupuncture is. No one knows how acupuncture was discovered; one theory is that a soldier was injured by an arrow and found that it stopped a pain that had been bothering him for years in another part of his body. The Chinese worked out a very complicated map of the human body showing exactly where to stick needles to relieve different ailments and pains. They even had maps for animals. The acupuncture examination is very challenging. A life-size bronze human statue with holes for the acupuncture points is covered with rice paper and the student has to stick the needle through the paper in exactly the right place!

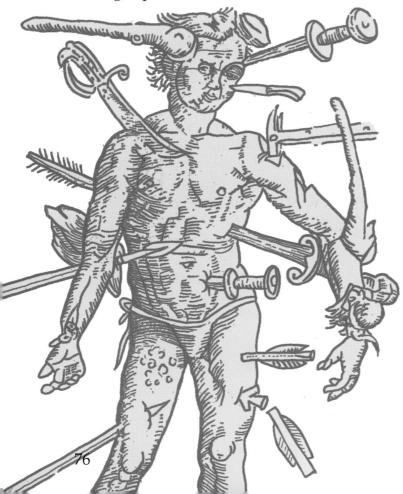

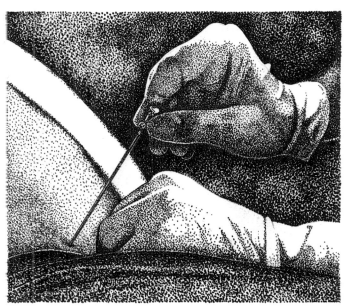

The Chinese thought that acupuncture worked by changing 'flows of energy' in the body. Western doctors who travelled to China could see that acupuncture sometimes worked, but they couldn't understand what the 'flows of energy' were. The nerve gate theory and the discovery of natural opiates offered an explanation. Acupuncture needles may stimulate the body to make natural painkillers and shut the 'gates' of the junction boxes in the spinal cord.

Surgery has been practised since the healing arts were first performed. Modern methods of anaesthesia make surgery safe and painless, but in the past it was agony for the patient and traumatic for the surgeon. Speed was essential. Over the centuries, powerful herbs like opium and mandrake root were used to induce sleep and relieve pain, but the results were often unreliable...

Take opium, mandragora and henbane in equal parts and mix with water. When you want to saw or cut a man, dip a rag in this and put it to his nostrils. He will sleep so deep that you may do what you wish.

A prescription from the 12th century...

There were no improvements until the late 18th century when many new chemicals and gases were discovered. One important property of each was its smell. Scientists were sometimes quite inebriated when they sniffed their discoveries! The gas nitrous oxide had such a powerful effect. that it was called 'laughing gas'.

London England 1799

Sir Humphrey Davy found the gas 'capable of destroying pain ...and may be used to advantage during surgical operations'.

Sniffing laughing gas or another newly discovered chemical called ether, became a fashionable way to make party guests jolly.

still in London 1815

The scientist Michael Faraday noticed that people hurt at 'ether frolics' did not feel pain, but still no one used the chemicals for surgery.

Georgia USA 1842

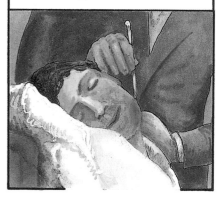

After an 'ether frolic', Dr Crawford Long used ether for painless surgery on a boy's neck, but he did not publicize his success.

Connecticut USA 1844

A dentist called Wells went to a demonstration of laughing gas. He realized the importance of its pain-destroying properties.

He used the gas while one of his own teeth was removed. When he awoke and had felt no pain, he immediately understood how important this was. Wells excitedly tried laughing gas on a few patients. But in a public exhibition, the patient cried out, although he felt no pain. Wells was discredited.

Boston October 1846   later the same year in London

William Morton continued the work using ether. He finally persuaded a surgeon to let him demonstrate.

The sceptical audience was impressed. They knew they had witnessed an important new discovery.

News travelled across the Atlantic with surprising speed. The surgeon Robert Liston used ether for his first painless operation, the amputation of a butler's leg.

Edinburgh Scotland 1847

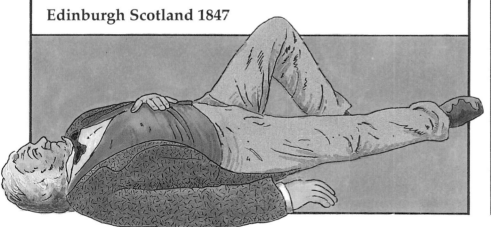

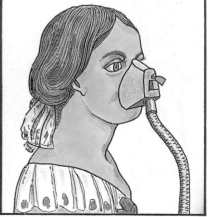

The professor of midwifery, Sir James Young Simpson, used to experiment on himself and his friends. One evening he tried a new chemical...When he finally woke up, he realized that this new substance, chloroform, was far stronger than ether. It rapidly became a popular anaesthetic.

Queen Victoria had chloroform during the birth of her 7th child Leopold in 1853. Anaesthesia was now used widely.

Meanwhile...

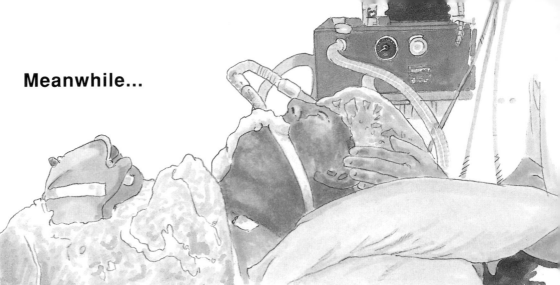

there were problems. Ether and chloroform were poisonous, explosive and caused horrible hangovers. In most countries, they are no longer used as anaesthetics. Nowadays a combination of drugs is used during surgery. These drugs not only put people to sleep, but stop pain and relax muscles, so that the surgeon's job is easier.

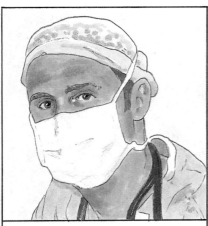

These are given by a doctor, an anaesthetist, who is with the patient during the operation and the first stages of recovery.

in the ward

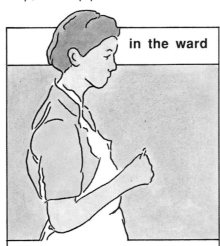

The first drug is called a pre-med. It may be derived from opium. It relaxes the patient and stops them feeling pain.

in the operating theatre

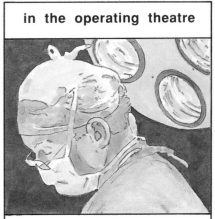

The patient is given an injection of an anaesthetic drug like thiopentone.

just 20 seconds later

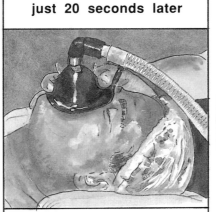

The patient is asleep and must be kept that way with anaesthetic gases. A mask is placed over the patient's face.

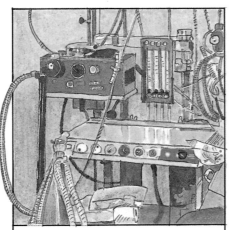

A mixture of oxygen, nitrous oxide and another anaesthetic gas, such as halothane, is then given.

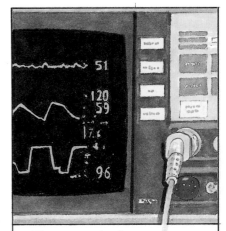

The anaesthetist monitors pulse, blood pressure and levels of oxygen in blood, constantly checking all is well.

A muscle relaxant may be injected, probably similar to curare, the 'active ingredient' of poisoned arrows!

Muscle relaxants are strong. The patient may stop breathing! No problem. The anaesthetist takes over.

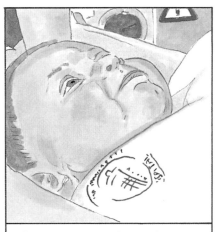

Once surgeons have done their work, anaesthetics are stopped and the patient will slowly regain consciousness.

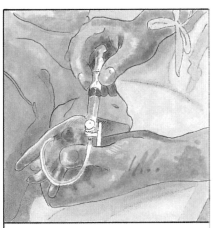

How do anaesthetics work? They do not have a common shape or size. They are quite different types of chemicals.

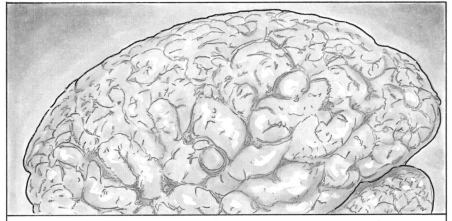

Scientists are now fairly certain that anaesthetics stick to proteins in nerve cell membranes, particularly in the brain.

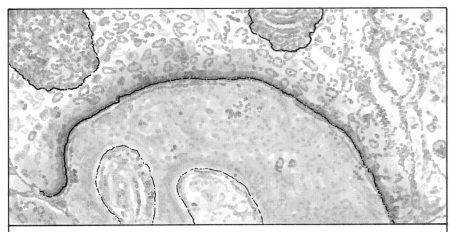

Some of them increase the power of one of the chemicals in the brain that reduces nerve signals. Their actions are reversible because they quickly dissolve out of the cell membranes.

Whatever the explanation, anaesthesia is now extremely safe. Suffering and disability can be relieved by complicated operations lasting many hours. But it is not always necessary to send the patient to sleep. For instance, at the dentist...

81

Having a tooth removed or filled is now a relatively painless exercise because of a woody shrub, the coca plant, that grows in the Andes mountains of South America. For over 5000 years the native Indians would mix its golden green leaves with lime. They placed the resulting paste between their teeth and gums as they went about their arduous work at high altitudes. The coca leaf paste reduced hunger and fatigue, but numbed the mouth.

In 1860 a European chemist called Albert Neimann extracted the active ingredient from leaves brought back from South America. The oily substance, which he called cocaine, had a bitter taste and numbed the tongue. Cocaine was used to 'revive flagging spirits'. The famous Sigmund Freud used it extensively, but people were already beginning to realize its dangerous addictive properties.

Freud had a friend, Carl Koller, who was an eye doctor. Koller was intrigued by the numbing qualities of the new drug. By dropping small quantities onto the eye of a patient he was able to perform painless surgery while the patient was still conscious. The idea of using cocaine as a local anaesthetic spread rapidly around Europe and America, but people were worried about using this addictive drug.

The first safer chemical substitute, procaine, was made in a laboratory in 1905, and many other similar chemicals were soon developed. Some could be applied directly to the skin, some were injected at a site near where pain relief was needed.

Local anaesthetics work by preventing nerve cells transmitting pain signals. Pain signals travel as minute electrical waves along nerve cells. These waves are triggered by sodium ions flooding in through the cell membrane of each nerve cell. Local anaesthetics stop the sodium ions flowing into the nerve cells. This completely stops the nerve signal and the sensation of pain. Most local anaesthetics are injected at a place where they can reach the finely branching nerves of the pain receptors.

As different nerve cells act in similar ways, the nerves that control muscle movement can also be blocked by local anaesthetics. This explains why face muscles can be paralysed after an injection at the dentist.

It is, of course, important that the effects of local anaesthetics are only temporary. The drugs are washed away into the blood stream and destroyed by enzymes in the body, particularly in the liver.

Mind Benders

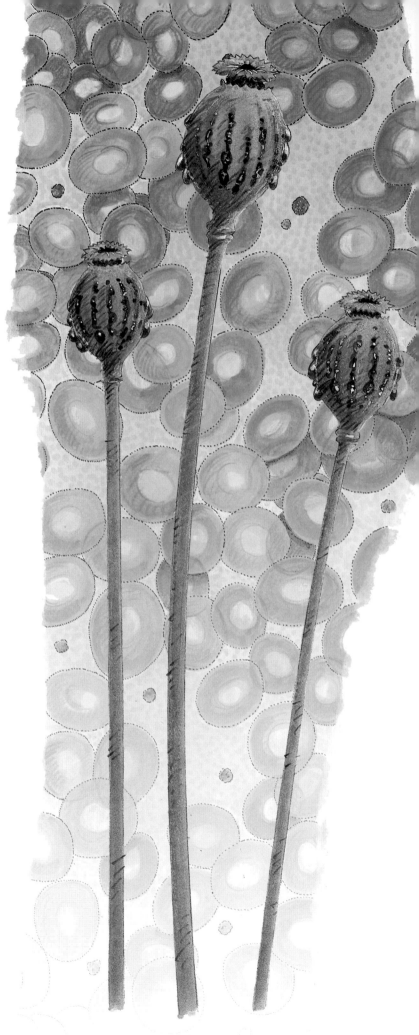

Throughout history, humans have sought ways to free their minds from the pressures of everyday existence, and experience dream-like states. Prayer and meditation, intense pleasure from art, music, and sexual passion, can all have powerful effects on mood and perception. But there are many drugs that can alter the mind. Their use goes back to the time of shamans and witches, maybe to the Stone Age and before. These drugs are called intoxicants.

There is no doubt of the power and danger of intoxicants. Societies older and wiser than ours used them sparingly, often only in sacred ceremonies. Now their use and abuse are widespread and lack religious significance. In previous centuries, all the drugs were crude plant extracts in which the action of the intoxicant was often tempered by other ingredients. Now pure active chemicals can be isolated and injected into the blood.

Some intoxicants are important medicines, particularly for relieving pain and anxiety. Milder intoxicants are used by millions of people who find them pleasurable and relaxing. Others use these drugs to improve concentration and ward off fatigue. But the most dangerous, powerful and addictive intoxicants have become a cause of much suffering, crime and disease.

Intoxicants are often divided into four groups depending on the main effects they have. Many affect people in several different ways which depend on the dose taken, and the manner in which they are used:

Hallucinogens cause strange visions. Most are extracted from plants like 'Magic mushrooms', the fly agaric toadstool, peyote cactus, *Cannabis sativa*, belladonna and henbane. One of the most powerful and dangerous hallucinogens, LSD, was made in a laboratory in the 1940s.

Inebriants generally make the user euphoric or drunk. The most common inebriant is alcohol, although chloroform, ether, benzene, butane and solvents have similar effects.

Hypnotics cause a state of sleep, calm and stupor. In early times, mandrake, kava and the opium poppy were frequently used as hypnotics. Now tranquillizer drugs and heroin (a chemical derived from opium) are common.

Stimulants result in increased mental and physical stimulation. Mild stimulants like tea, coffee and cocoa are a common and harmless part of everyday life although their use was once restricted and ceremonial. More dangerous and addictive stimulants are nicotine, cocaine, amphetamines and 'Ecstasy'.

All intoxicants act on chemicals in the brain. The ten billion nerve cells in the brain have long projections called axons that end with many branches. Each branch may connect with a different nerve cell. Electrical signals are rapidly carried down axons, but, when the signal comes to a nerve ending, it is transmitted across the gap to the next nerve cell. This happens by the rapid release of chemicals called neurotransmitters. Neurotransmitters briefly bind to receptor molecules on the adjacent nerve cell, as a key fits in a lock. All the 'mind bending' drugs affect neurotransmitters in some way.

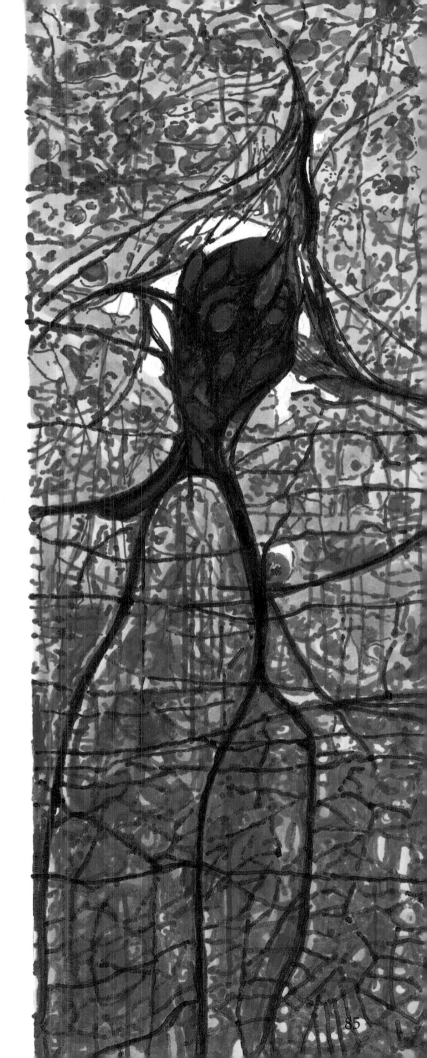

Stone Age hunter gatherers cultivated a bushy plant we know as *Cannabis sativa* for its tough fibres (hemp). Hemp made strong ropes and sacking, but as the tribes harvested their crop, they may have discovered another use for the plant. Archeologists excavating an ancient site in Romania from 3000 BC found a 'pipe-cup' containing charred hemp seeds. Had these early humans already discovered the intoxicants in cannabis leaves and resin?

Cannabis sativa yields one of the most commonly used intoxicants in the world. Its powerful resin, compressed into blocks for smoking or adding to food, is known by many names such as hash, hashish, dope, pot, ganga. Its leaves are smoked as marijuana or grass.

The Rastafarians of Ethiopia consider the drug sacred. Many ancient cultures used it as a medicine, particularly the Chinese. Cannabis was not brought to the West in any great quantity until the mid-19th century when it was used to treat many ills. Queen Victoria's doctor, for instance, was a great enthusiast. In the 1950s and '60s, the jazz clubs of the Southern USA were the setting for a widespread increase in popularity of the drug.

By this time, however, the drug was illegal. Its intoxicating properties led to its being banned in the UK in 1928. It was still occasionally used as a medicine until 1973, until that too became illegal. Cannabis was banned in the USA in 1937, although some American states now regard possession as a trivial offence, equivalent to a minor traffic violation.

Cannabis most commonly causes mild euphoria, relaxation and a feeling of well-being. Senses are sharpened, the pulse rate rises, blood pressure drops, appetite increases. Under the influence of cannabis, short-term memory and motor co-ordination are impaired. It is not, for instance, safe to drive a car. Cannabis may sometimes cause depression, anxiety and mild hallucinations. In rare cases, it may unmask hidden mental illness. Cannabis is not strongly addictive, but is harmful because it also contains tar and other cancer-causing chemicals.

The active ingredients of cannabis are chemicals called cannabinoids. It is not entirely clear how cannabinoids affect the brain. They may act like mild anaesthetics, dissolving in cell membranes and temporarily altering the behaviour of nerve cells.

But recently scientists have found molecules in brain cell membranes that can attach cannabinoids as a key fits a lock. These cannabinoid 'receptors' are also found in other parts of the body. Why should they be there? The intriguing possibility is that our bodies may make molecules like cannabinoids that could influence, for instance, our moods and perception of pain.

Some people believe that cannabis should be legalized, that it is no worse than tobacco or alcohol. They argue that the law is ineffective and encourages crime. Others believe that cannabis is a useful medicine. Patients with serious illnesses such as multiple sclerosis and cancer sometimes find it relieves their symptoms. The opposing opinion is that the law should not be changed; that cannabis is harmful, and its use might encourage experiments with more dangerous drugs.

One of the most dangerous hallucinogenic drugs is not extracted from plants, but was discovered by chance in 1943. This drug became a central part of the psychedelic 'hippy' culture of the 1960s. It is called lysergic acid diethylamide, or LSD.

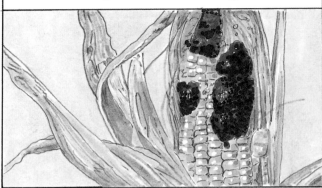

Europe The Middle Ages

Claviceps purpurea is a poisonous fungus that grows on wheat. Contaminated wheat causes gangrene and convulsions, but the fungus, surprisingly, was a useful medicine for inducing childbirth.

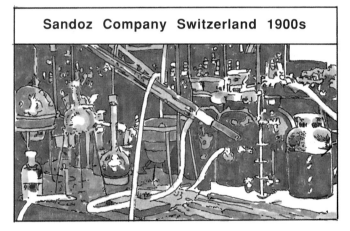

Sandoz Company Switzerland 1900s

Scientists found that active chemicals from the fungus contained lysergic acid. They changed this chemically to make useful drugs for treating migraine and improving mental function in the elderly.

still in Switzerland 1940s

LSD25

The 25th chemical derivative of lysergic acid, code-named LSD-25, was made by Albert Hofmann at Sandoz. Scientists could find no use for it. It should have been destroyed.

But, in April 1943, Hofmann had a hunch that it had undiscovered properties. While he was purifying the crystals once again, he was overwhelmed by a strange dreamlike state and fantastic colourful visions. He thought this was caused by the LSD-25. So three days later he deliberately took a minute dose (less than 1/1000 gram).
He later wrote:

`'Pieces of furniture assumed grotesque threatening forms...the lady next door...was no longer Mrs R but a malevolent insidious witch...I was seized by a dreadful fear of going insane...I was taken to another world, another place, another time...Was I dying?'`

LSD causes profound changes in perception of the senses. Touch can be heard, sounds can be seen. Sense of time is distorted and sense of self disintegrates, sometimes causing fear and panic.

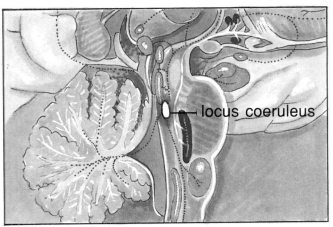

locus coeruleus

LSD seems to affect the 3000 cells of the locus coeruleus in the brain. These cells send axons all over the brain, connecting with many millions more nerve cells on their way.

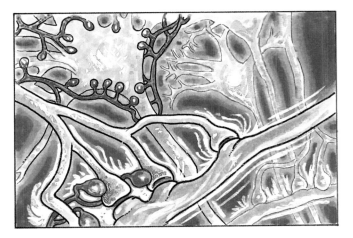

These few cells normally 'funnel' input from nerves responding to sight, sound, touch, taste and smell. After LSD is taken they become hyperactive and amplify the signals from the sensory nerves.

The effects of LSD usually wear off without causing any long-term harm, although users may feel confused, disorientated and suffer 'flashbacks'. The danger of LSD and other hallucinogens is that a single dose may unmask a hidden mental illness and leave the user permanently scarred. Another problem is that the illusions may cause such panic and fear that people become violent and may even commit murder or suicide. When LSD was first discovered, scientists and doctors conducted many trials to find a medical use for this powerful drug.

But its effects were unpredictable and often terrifying to sick patients. In the drug-culture of the 1960s, LSD became increasingly popular. Unscrupulous chemists began to make it illegally. As little as one-millionth of a gram of LSD, sold on small squares of blotting paper that dissolve on the tongue, could provoke powerful visions of good or evil and alter states of consciousness for several hours. The number of tragic deaths and injuries increased. Laws were passed imposing powerful restrictions on the use of LSD, and harsh penalties on manufacturers and suppliers.

Microscopic yeasts feed on plant sugars to give us the most common intoxicant of the Western world. It is probably the oldest drug to be used and abused by our species. At its worst, the drug can be deadly, damaging, degrading and addictive. But it is also used for celebrating special occasions, relaxation, and stimulating appetites for food and sex. It may even be beneficial to health if taken in small quantities. The drug is, of course, alcohol. Its proper chemical names are ethanol or ethyl alcohol.

All alcoholic drinks have their origins in fermentation. This is a chemical process in which yeasts feed on sugars and starches in plants, making ethanol and carbon dioxide as waste products. The yeasts grow until the alcohol concentration reaches about 14% and then the yeast cells die. Alcohol can be made from any plant, but wine is usually made from grapes, and beer from barley (with hops added for flavour). The amount of alcohol in a drink can be increased by distillation. The liquor is gently heated and the alcohol evaporates first, ahead of the other liquids in the brew. This gives us spirits like brandy, whisky and rum.

Alcoholic drinks don't just contain ethanol, they contain other chemicals like tannins (also found in tea), esters (that smell like pear drops) and aldehydes, which can be converted into acids. All these chemicals add flavour to the drink. For instance, the distinctive taste of rum is due to a chemical called 2-ethyl 3-methyl butyric acid (imagine someone asking for that in a bar!). Casks and barrels that store wines, beers and spirits add yet more flavour-giving chemicals.

Early humans probably discovered alcohol by accident when they ate fermenting fruit. Sumerian doctors wrote prescriptions for beer on clay tablets in 2000 BC. The Hindu Ayurveda, written in 1000 BC, describes in detail alcoholic drinks and the consequences of intoxication. Greek and Roman myths tell many stories of drinking and drunkenness, and wine has great significance to both the Hebrew and Christian religions. In Mediterranean countries, wine is considered a basic food, along with bread, cheese and olive oil.

But the dangers of alcohol were also recognized by the ancient societies. By the 7th century, Islam dictated a total prohibition of alcohol which still exists today. The Qur'an condemned wine drinking, and the disciples of Muhammad imposed this rule on all lands they conquered. Buddhists, Hindu Brahmins and some Protestant sects have also banned alcohol from their lives. British alcohol drinking habits first caused concern in the 18th century. In 1726, a petition to parliament from eminent doctors talked of the *'...fatal effect of frequent use of several sorts of spiritous liquors upon great numbers of both sexes, rendering them diseased, not fit for business, poor, a burthen to themselves and neighbours...'*

Alcohol consumption in the UK rises and falls, with changes in the fortunes of its people, and the laws controlling its sale. At any one time, the mental and physical health of at least 300,000 people is severely damaged. Alcohol is also a cause of many crimes, accidents and much domestic violence and unhappiness.

Make not thyself helpless in drinking in the beer shop. For will not the words of thy report repeated slip out from thy mouth without thy knowing that thou hast uttered them? Falling down thy limbs will be broken, and no one will give thee a hand to help thee up. As for thy companions in the swilling of beer, they will get up and say, 'Outside with this drunkard'.

translation of hieroglyphs from the Precepts of Ani
c1500 BC advising moderation in drink

Alcohol is a simple chemical that is readily absorbed through the stomach and intestines, straight into the blood. About 95% is broken down in the liver. The rest is excreted by the lungs (hence the 'breathalyser' test for drunk driving) and the kidneys. Alcohol widens tiny blood vessels in the skin and brain, increases stomach and salivary secretions, and raises blood pressure. It makes people dizzy because of changes in the fluid that normally controls balance in the inner ear. Alcohol also alters the water balance of the body. But the most important effects of alcohol are on nerve cells, particularly in the brain. Nerve cell membranes are changed and their electrical messages altered.

These are short-term effects, although the 'hangover' of the following day is not a pleasant experience. Tragically, some people (alcoholics) become dependent on alcohol. Alcohol can cause permanent damage to the liver and brain of people who regularly drink dangerous quantities. The exact quantities that can damage health vary from person to person. Most doctors and scientists estimate that the upper limit of safe drinking for female adults is 14 units a week and 21 units a week for men. Women seem to be more susceptible to alcohol than men because they are usually smaller and lighter than men, have less water, and slightly more fat in their bodies.

The concentration of alcohol in drinks varies between 2.5% (weak beer) and 55% (strong spirits). A glass of wine, a measure of spirits or a half-pint/quarter-litre of beer all contain 8–12 grams (g) of pure ethanol. This amount is also called one unit.

Two units of alcohol give a blood level of about 30 milligrams (mg) of alcohol per 100 millilitres (ml) of blood. This may have little effect, but it can make people feel relaxed, excited and elated.

After drinking 5 units of alcohol (40 g), people may become clumsy, emotional and aggressive. Their blood alcohol level will be about 80 mg per 100 ml blood. They are four times more likely to cause an accident if they drive a car.

The levels of alcohol in the blood rise as more is consumed. Severe intoxication occurs in 90% of people who have 150 mg of alcohol per 100 ml of blood. This is usually the result of drinking 10 or 12 units of alcohol over a few hours. At this level, people are 25 times more likely to have a car accident.

People who manage to drink in excess of 20 units on one occasion risk more than a car accident. A blood level of 300 mg alcohol per 100 ml causes a coma. Blood levels above 400 mg per 100 ml are usually fatal. At this blood concentration, alcohol paralyses the nerves that control breathing.

Alcohol is more powerful and dangerous to children, young adults, and to anyone who drinks it very rarely. As little as three units could kill a two-year old child, for instance.

Humans seem to be continually discovering new intoxicants in the world around them. Many chemicals that have been developed in the 20th century for industry or households are abused. In the 1960s, there was a fashion among American teenagers to get 'high' by sniffing petrol (gasoline). Then, in the early '70s, the craze was 'glue sniffing'. Now, in the Western world, a small proportion of teenagers experiment (often only once because it can be extremely unpleasant) with solvents.

Many common household goods contain volatile inebriating solvents. They range from aerosol sprays to butane gas, dry-cleaning fluids, paint and paint thinners, correcting fluids and petrol. The effects of breathing fumes, or spraying aerosols into the mouth, are rather similar to alcohol. All these chemicals are thought to act like anaesthetics, sticking to proteins in nerve cell membranes. They may also alter some of the chemical messages (neurotransmitters) in the brain.

The inebriating solvents do not usually cause much long-term harm to the body, but there are rather a lot of reasons why they are dangerous. Here are some of them:

Solvents can cause frightening hallucinations that place the user in danger.

Solvents may contain poisonous substances like lead.

Combining solvents with alcohol or other drugs is particularly risky.

Solvents sprayed into the mouth can make the throat swell and suffocate the user.

Solvents can make people unconscious.

They are often inflammable. Users risk fire or explosion.

Solvents can affect the heart. Under their influence, sudden fright or exertion can cause a fatal heart attack.

Every year in the UK about 100 young adults die needlessly from solvent abuse. Often it is their first 'experiment'.

The opium poppy (*Papaver somniferum*) has been a source of powerful medicines since Neolithic man built lakeside villages. In ancient Egypt, Assyria, Greece and Rome, the medicines of the poppy relieved pain, calmed anxiety and helped the breathless. The poppy gives us the most potent painkillers ever discovered, but also drugs that cause endless suffering, torment and pain.

Opium has always been an important drug. By the 16th century, doctors wrote that stopping the drug after long-term use led to *'great and intolerable distresses, anxieties and depression of the spirit...'*

Opium is a rubbery substance that congeals from the sap of the unripe poppy capsule. It is extracted by cutting fine slits in the outer surface of the capsule and collecting the juice the following day.

In 1805 a German chemist, Friedrich Seturner, purified an alkaline chemical from opium. He named it morphium (morphine) after Morpheus, the Greek god of dreams. Morphine was a powerful painkiller.

By the American Civil War (1861–1865) the hypodermic syringe had been invented. This meant that morphine could be given quickly and accurately to wounded soldiers. But many returned home addicted to injectable morphine.

The recreational use of opium and morphine really began in Europe in the 19th century and was encouraged by the British Romantic authors.

In his essay, 'Confessions of an English Opium Eater', Thomas De Quincey wrote ...*'Here was the secret of happiness...might be bought for a penny, and carried in the waist coat pocket; portable ecstasies might be corked up in a pint bottle...'*

Morphine and opium were also included in a number of 19th century medicines and tonics prescribed by doctors or sold by pharmacists. Many people became addicted this way.

Famous poetry was written under the poppy's influence, but the dangers of addiction soon became clear. Opium dens, where addicts would smoke opium for days on end, were widespread.

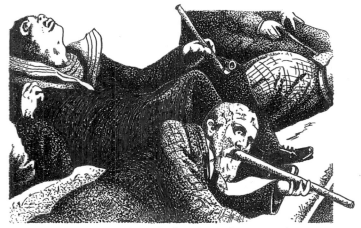

Morphine and other similar drugs became the scourge of the 19th century, but until 1912 there was no attempt to control the traffic and supply of these dangerous drugs.

In 1875 a new chemical derivative of morphine was made. It was considered a non-addictive substitute for morphine and was sold for a time by the Bayer drug company as a cough medicine.

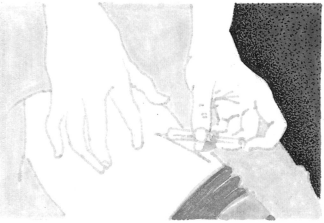

This drug was called heroin and its abuse has plagued our planet in the 20th century. It is a major cause of crime, violent death, disease and suffering throughout the world.

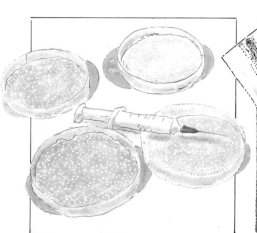

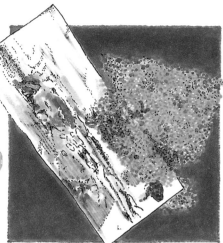

Heroin can be injected, sniffed, or heated and its fumes inhaled. It can cause drowsiness, euphoria, and feelings of contentment.

First-time users may suffer severe nausea and vomiting, and may not try it again. If they do, they may easily become addicted.

Tolerance develops first. The user needs more and more heroin to produce the same effect.

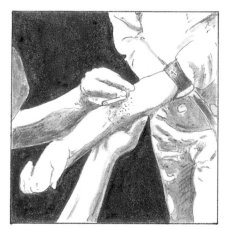

Next comes physical dependence. Stopping heroin provokes unpleasant and painful withdrawal symptoms.

Then the addict becomes a compulsive drug seeker whose only aim is to find the next dose – at any cost.

Injecting heroin means shared syringes and shared diseases, especially hepatitis and AIDS.

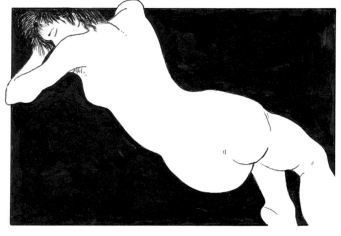

The cost of addiction to society, in terms of crime alone, runs into billions of pounds/dollars. The cost to the addicts and their families is incalculable. But why are opiates so powerful?

Addicts are malnourished and suffer constipation. Overdoses of heroin kill because the nerves that control breathing are paralysed.

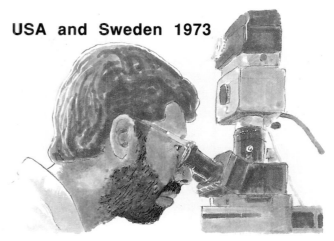

Opiates are extremely powerful but their effects are rapidly reversed by certain chemicals. This made scientists think there might be molecules on cells that would strongly bind to opiates.

Scientists in three different laboratories finally identified opiate receptors in the brain. They were mostly concentrated in the areas that control perception of pain, the senses and breathing.

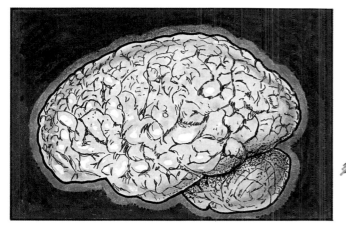

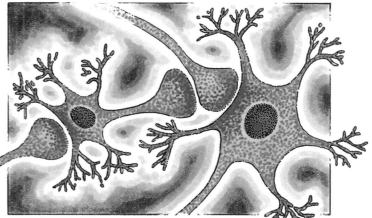

But why were the opiate receptors there? Maybe natural opiates existed in the brain. Scientists isolated two small proteins that acted like morphine.

Natural opiates are called endorphins. They are made by nerve cells and control the sensation of pain in the body.

Morphine, heroin and other similar molecules are still important medicines for people with severe pain. Now that doctors and scientists realize their power, their use is carefully controlled. The more we understand about the opiates and their natural relatives, the greater the chance of creating safe and powerful painkillers and removing the scourge of addiction.

Scientists hoped that these might form the basis for new, powerful, but non-addictive painkillers, but those so far developed are no better than morphine or heroin, and are still addictive.

Not all intoxicants are dangerous and addictive. Some mild stimulants are widely accepted. Most of them are made into popular drinks, and there is little evidence that they are harmful or seriously addictive.

Coffea arabica is a shrub or small tree. It originally came from the highlands of Ethiopia. In the 15th century, Arabians first made a drink from the two seeds (beans) inside each berry. Coffee was brought to England in 1601 and coffee houses became popular throughout Europe.

The tea plant (*Camillia thea* or *Thea sinensis*) grows in tropical hill country as a shrub or tree. It is native to China, India and Japan. The leaves, large (Indian-type) or narrow (China), are fired after harvesting. Tea was first popularized in England in the 19th century, some doctors first thinking it a 'deadly poison'.

The cocoa tree (*Theobroma cacao*) comes from the Amazon valley and was brought to Mexico over 500 years ago. The Aztecs called it chocolatl (food of the gods) and served it in goblets of gold. The Spanish invaders took some back home and by the 1650s much of Europe was enjoying drinks or bars made from the roasted de-husked beans, sweetened with sugar.

Cola is also a stimulating drink. Originally the stimulants came from African cola nuts. Some 'cola' drink manufacturers still use them in their drinks. Cola nuts are popular in West Africa, particularly with Muslims, whose religion forbids the use of other intoxicants.

Tea, coffee, cola nuts, cocoa and chocolate bars all contain stimulants from a group of chemicals called methylxanthines (mee-thile-zan-theens). You may not have heard this complicated name before, but you probably know the name of the most common methylxanthine, caffeine. In countries where tea and coffee are popular, each person consumes, on average, one-fifth of a gram (200 mg) of caffeine per day. There is about twice as much caffeine (115 mg) in a cup of roasted coffee as there is in instant coffee. A cup of tea contains between 9 and 50 mg depending on how long it is left to brew.

Methylxanthines stimulate nerve cells and cardiac muscles. They cause mild relaxation of smooth muscles, slightly increase the rate at which the kidneys filter blood, reduce tiredness and improve concentration. Caffeine decreases reaction time, and increases calculation speed (although it does not improve accuracy!). Typing, driving and other motor skills are also enhanced, particularly if people are tired. This has been proved in scientific tests using doses found in one to three drinks.

There is little evidence that moderate consumption of these stimulants is harmful to health or that they are seriously addictive. (Experts even say that chocolate is non-addictive although many people would disagree!) Caffeine in higher doses (over four-fifths of a gram per day) may cause anxiety and stomach problems, and someone who changes from a high daily consumption to low or no caffeine, may experience temporary discomfort.

However, there are more powerful and harmful stimulants that cast a shadow over our society...

For example, there is one addictive stimulant that can be brought legally at shops, restaurants and bars all around the world. The drug is nicotine and it is usually obtained from the tobacco plant (*Nicotiana tabacum*) in the form of cigarettes. Tobacco is now the single biggest cause of early death in the developed world.

When Christopher Columbus arrived in North America, he found the native Indians smoking tobacco in pipes or small rolls of dried leaves. Soon after, two British sea captains took three of these Indian natives, and an ample supply of tobacco, on a voyage to London. During the journey, many sailors tried the tobacco. Within a few years, the explorers of the high seas had established fields of tobacco plants at their ports of call in Europe, the Americas and Africa. As the local inhabitants began to smoke tobacco as well, the fields became plantations. By the 19th century tobacco was grown and smoked all over the world. In most societies, there was initial opposition to this new habit. Many people thought it enjoyable, distinguished and rather fashionable to smoke, but others felt it to be harmful and dangerous. Most countries first tried to ban the drug, but that created a black market and encouraged smuggling.
The compromise usually reached was to legalize tobacco but impose high taxes on the buyers.

Tobacco smoking was at its height in the Western world in the early and mid-20th century (at one point over 80% of men in the UK smoked), but it is declining now. For instance, in 1991 26% of people in the USA regularly smoked cigarettes and the figures for the UK in 1992 were 29% of men and 28% of women.

The tip of the burning cone at the centre of a cigarette reaches a temperature of 1093°C. It is a tiny blast furnace, a miniature chemical plant that uses hundreds of the 4000 or more chemicals in a cigarette to create many more. Some of the most dangerous chemicals are not present in the unlit cigarette. Three chemicals that are most important are tar, nicotine and carbon monoxide.

The **tar** in cigarettes, like the black sticky substance used to surface roads, is made when organic matter is burned at high temperatures. Minute particles of tar are the most damaging part of cigarette smoke and contain many cancer-causing and irritating chemicals. As they are less than one-thousandth of a millimetre across, they become lodged in the tiny alveoli (air sacs) of the lung.

The tar droplets also carry the highly addictive stimulant **nicotine** which is found in the leaves of the tobacco plant. Once inhaled, nicotine passes very rapidly into the blood carried from the lungs, reaching the nerve cells in the brain less than ten seconds after each puff of the cigarette.

Carbon monoxide is a gas made when organic chemicals in cigarettes are burned. This poisonous gas is, for instance, a major pollutant in car exhaust fumes. Like nicotine, carbon monoxide passes easily into the blood stream of the lungs where it combines with the oxygen-carrying molecule, haemoglobin, in red blood cells. Haemoglobin usually carries the waste product carbon **di**oxide back to the lungs, where it is breathed out and replaced with oxygen. Carbon **mon**oxide binds more tightly to haemoglobin. If blood levels of carbon monoxide rise above a certain level, a serious shortage of oxygen can occur.

While smokers may enjoy the effects of nicotine, many of the other chemicals in cigarette smoke are harmful to health. For instance, some of the chemicals in tar damage DNA. Over time, this damage leads to cells that are out of control – cancer cells. Smokers are eleven times more likely to develop lung cancer than non-smokers. Smoking cigarettes can also cause cancer of the mouth, pharynx, larynx, oesophagus, stomach, pancreas, kidney and bladder. In fact, smoking doubles the risk of contracting cancer overall. Chemicals in cigarette smoke also damage cells lining blood vessels, thus affecting blood clotting, and damage the muscles of the heart. This means that smoking increases the risk of heart attacks, strokes, and brain haemorrhage.

Every day in Great Britain 450 children start smoking for the first time. The reasons why people, especially young adults, start smoking are varied and often quite complicated to unravel. The reason why people cannot stop smoking is simple. They are addicted to nicotine.

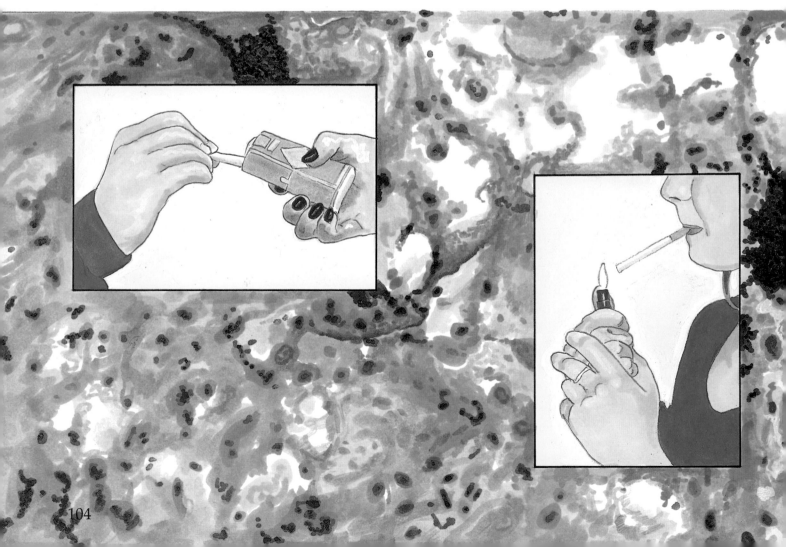

Nicotine is a simple chemical, but it has a number of quite complicated effects on the brain and other parts of the body. The nicotine molecule has a similar structure to one of the chemicals used by nerve cells to transmit their messages. This chemical is called acetylcholine. Once inhaled, nicotine rapidly reaches the nerve cells and mimics the actions of acetylcholine to stimulate new nerve impulses. But, because nicotine is not released so rapidly from cells, it can then inhibit the next wave of nerve impulses. This happens especially in regions that control physical and mental arousal, learning, memory, and some emotions. Nicotine also binds to cells in the adrenal glands above the kidneys. This causes the hormones noradrenaline and adrenaline to be released into the bloodstream.

These changes, and others, in chemicals and cells, may make the smoker slightly more alert when drowsy, and calmer when tense. The effects are usually quite small and, with regular smoking, are less noticeable. This is because the body cells become 'tolerant' to nicotine. More and more is required to achieve the same effect. The smoker is now in the first stage of addiction, and quite rapidly becomes 'dependent' on nicotine, craving for a cigarette if supplies run out, sometimes experiencing quite uncomfortable withdrawal symptoms. In fact, heroin, cocaine and alcohol addicts often report that it is harder to give up smoking than their other problem drug. Scientists and doctors are still uncertain why people need nicotine so much. One popular theory is that people smoke cigarettes to avoid the unpleasantness of **not** smoking!

A slice cut through a smoker's lung showing black tar particles from the smoke. Defender cells attack the tar and fibrous tissue forms because of the irritation.

In the mountainous regions of Peru, Indians of the Inca civilization believed that angels from their Sun God had given them a most wonderful gift, leaves from the coca bush. Chewing these leaves *'...satisfied their hunger, provided the weary and fainting with new vigor and caused the unhappy to forget their miseries'*. At first only royalty and priests used them, but, later on, commoners, especially those who worked at high altitudes in gold mines, exploited the power of the Sun God's gift. There is little evidence that coca leaves were abused by the Incas, or caused them any harm, but they were exploited by conquering Spanish invaders. The Spaniards amassed vast fortunes of gold at very little expense by giving coca leaves to their Inca slaves. The slaves would then need very little food to work extremely hard.

These leaves were not brought back to Europe in any quantity until the mid-19th century. In 1860, Albert Neimann, a German chemist, purified the active ingredient and called it cocaine. Europeans were fascinated by coca leaves and cocaine. Angelo Mariani, a Corsican chemist, patented a mixture of coca extract and wine called 'Vin Mariani', soon recommended by doctors for many illnesses. The Pope even gave him a medal! This inspired John Pemberton, a pharmacist from Georgia USA to combine coca extracts and wine as a new headache remedy. He began to sell 'Coca Cola' in 1886. It was an immediate success, and carried on selling even when he substituted wine with a cola nut extract containing caffeine. By 1888, 'Coca Cola' was made fizzy with soda water. You can now buy 'Coca Cola' all over the world. There is only one difference – it contains no cocaine. Why?…

The problems began with the father of psychoanalysis, Sigmund Freud. He read of Peruvians and coca leaves and started experimenting with pure cocaine, first on himself! He was very enthusiastic about its effects, writing an influential article about how it might help patients suffering from anxiety and depression. He also noticed that it could numb the skin and mouth. Then tragedy struck. Freud had a close friend who had become addicted to morphine after his thumb had been amputated. Freud substituted cocaine for morphine. At first it worked, his friend no longer craved morphine, but soon he needed larger and larger doses of cocaine until he became dangerously deranged. All over Europe, doctors who followed Freud's recommendation found their patients would experience wonderful but brief euphoria followed by a 'crash' into severe depression.

It was not long before one eminent German doctor pronounced cocaine the 'third scourge of mankind', in league with morphine and alcohol.

The numbing action of cocaine was briefly used for local surgery, part-icularly to the eye, but as its danger was appreciated, chemical derivatives such as procaine were developed, that numbed the nerves without causing addiction.

Chemical purification of the active ingredient of the harmless, but stimulating, coca leaves had led to a highly dangerous and expensive drug that was soon exploited by criminal organizations. In the USA alone, an estimated four million adults now spend 39 billion dollars a year abusing cocaine-based drugs.

Cocaine is a white powder that is usually laid out on a flat surface in lines and sniffed through a straw up the nose. It can also be dissolved in water and injected. Crack, or freebase, is the most powerful and addictive form. Its white raisin-sized crystals can be smoked, or its vapour inhaled after burning. Users of cocaine experience euphoria, increased alertness, energy, physical power and mental clarity. They feel little hunger. The effects are not dissimilar from those obtained from chewing coca leaves, but much more powerful.

Sometimes the effects of cocaine can be alarming. Users may hear strange sounds and voices, and suffer from delusions, fearing they are surrounded by enemies out to destroy them. They may feel unpleasant sensations in their skin and believe that they are infested with lice or worms. On a 'bad trip' the user may have symptoms similar to someone who suffers from the mental illness schizophrenia. Cocaine sniffing can also cause severe damage to cells inside the nose.

There are other stimulants that have similar effects to cocaine. They are usually related to a drug called amphetamine.

Since the early 1900s, one of the best treatments for asthma was a hormone called adrenaline, but it was difficult to administer. A pharmacist working for the Liley drug company learned that a plant called *Ma huang* had been used to treat asthma in China. He extracted the active ingredient, which he called ephedrine. *Ma huang* was a rare plant, a synthetic substitute was needed. By the 1930s one was found, a chemical named amphetamine. This was marketed as an inhaler, and was sold without a prescription all over the world. But some people found that the drug could be a powerful stimulant, particularly if the inhalers were broken open and the contents swallowed. Students used amphetamine at exams, German pilots used it on bombing missions, and British, American and Japanese soldiers used it to keep alert in battle.

By 1948, 55% of all young Japanese were dependent on the drug. In the 1960s, the hippies of San Francisco mixed amphetamines with LSD, often with disastrous effects. Soldiers in the Vietnam war relied on 'purple heart' amphetamine pills. Amphetamine is now commonly sold in wraps of white powder, or as pills, and is called 'speed', 'whizz', 'sulph', 'glass' or 'crystal', etc.

'Ecstasy', 'E', is not so addictive as amphetamine, but causes a similar euphoria, benevolence, and a feeling of connection with anyone close by. But 'E' has hidden dangers. Apart from nausea, diarrhoea, sweating and a dry mouth, high doses cause severe anxiety and an inability to sleep. Some people have inherited a mild genetic defect from their parents which means that their body cannot break down the drug properly. Ecstasy then becomes a very dangerous drug causing severe side-effects. It can even kill.

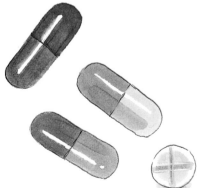

Because they have very similar effects, scientists think that cocaine, amphetamines and 'Ecstasy' all work in much the same way. They affect the chemicals made by a few nerve cells in a part of the brain stem called the locus coeruleus that are especially involved in emotions. Of the billions of nerve cells in the brain and spinal cord, only 3000 are found in the locus coeruleus, but their long thin axons reach for vast distances all over the brain, branching and branching so that they touch as many as one third of all the other nerve cells there.

All the 'mind benders' we have described interfere with chemical neurotransmitters in the brain.

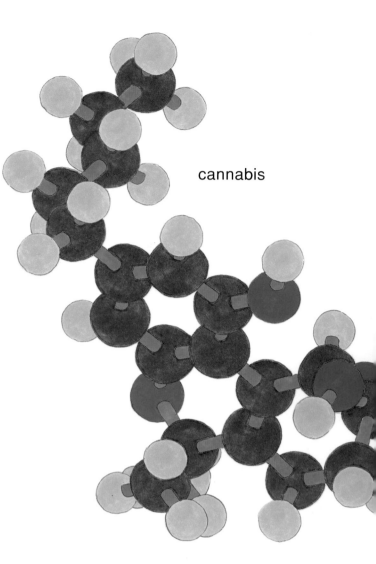

cannabis

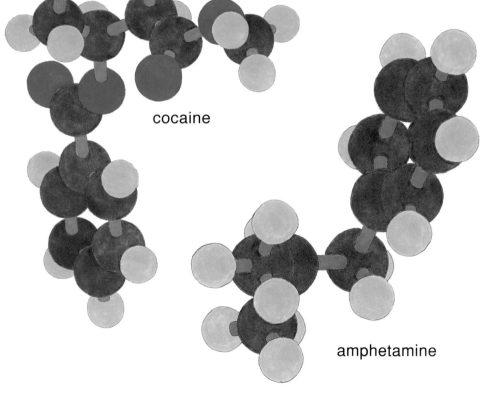

cocaine

amphetamine

morphine

LSD

Amphetamine, for instance, is very similar to the neurotransmitter and hormone, noradrenaline.

Alcohol increases the activity of an inhibitory neurotransmitter called gamma-amino butyric acid or GABA. As 25–40% of all the gaps (or synapses) between nerve cells use GABA, you can understand how alcohol can affect people so much!

Opiates resemble neurotransmitters called endorphins which are natural opiates made in the brain.

The chemicals in cannabis bind to receptors in synapses but scientists have not yet identified the neurotransmitters they mimic.

LSD resembles neurotransmitters called serotonin and noradrenaline.

The structure of the nicotine molecule is very similar to the neurotransmitter acetylcholine.

Understanding the way intoxicants alter behaviour and affect chemicals of the brain, has helped scientists learn about the way the normal brain functions. It has also helped them design drugs like tranquillizers and anti-depressants that can help the mentally ill. These drugs also change the chemical messages of the brain, but that is another story!

alcohol

Potion Power

Since the First World War, human health has improved dramatically in the richer countries of the world. Of every 1000 children born in 1920, about 80 died before their fifteenth birthday, most from infections in the first year of life. Nowadays only 4 children in 1000 would be expected to die, mainly from accidents, cancer or inherited diseases.

Better food and housing, vaccination, and hygienic living conditions have all helped. Drugs like antibiotics have saved millions of children and relieved the suffering of many more. Drugs help during childbirth, and save babies born too early. Some children and adults with cancer can be cured by drugs, and those with diabetes, epilepsy and other serious diseases, can lead normal lives. Irritating and debilitating allergies are readily eased. Pain relief is now a sophisticated science. Other medicines, especially for diseases of the heart and blood vessels, can help people as they grow old. But new and better drugs are still needed. There are many illnesses for which there is no cure, and some drugs need improving.

Discovering and developing new medicines takes years of dedicated scientific research, inspiration, luck and a great deal of money. As people live longer and healthier lives, new drugs must, above all, be safe.

A potential drug is first tested on animals. Then it is given to healthy human volunteers who are paid to test the new medicine. Doses are carefully increased and the volunteers are monitored for side-effects. If the drug still seems safe, it can then be given to sick patients. Often the drug is tested alongside a 'dummy' pill (placebo) which looks and tastes the same, but contains no drug. (This is because people often feel better if they are given a tablet with no drug at all!) Even the doctors do not know who has received the real or the dummy drug until the trial has ended. If the drug is proved useful, the Government controller of medicines is given all the scientific information and asked to license the new medicine.

This all takes about 10 years, £150,000,000, and the efforts of 300–400 people. It is not surprising that only 1 in 10,000 potential drugs ever gets this far!

This all may sound like good news...

...but doctors and scientists have had to learn from past mistakes. Many innocent people, or their unborn babies, have been harmed by unexpected side-effects of new medicines. Even safe and useful drugs can be dangerous if taken in excess, or used incorrectly. For instance, the major problem with antibiotics is that bacteria readily become resistant to them. Yet antibiotics are sometimes taken for virus infections, and patients often fail to follow their doctors' instructions. They do not finish the complete course, they hoard their tablets and even share them. All this encourages the bacteria to become resistant.

Most companies that make medicines are responsible and extremely careful, but the laws controlling drug safety and quality vary around the world. Medicines are becoming more and more expensive, especially as people live longer. Many of the drugs that we find so useful are too costly for those who live in the poorer nations.

Some drugs prescribed by doctors can be addictive if they are not carefully monitored. The drug nicotine is legally sold all over the world in cigarettes, cigars, snuff and chewing tobacco. Illnesses caused by tobacco are now the major cause of death of middle-aged people in many parts of the world. As the number of people using tobacco declines in the richer countries, more and more people are taking up this habit in the Third world. Alcohol, a legal drug in many countries, takes its toll on both the abusers and their families.

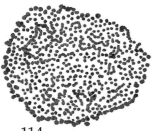

At some stage in history, many of the most dangerous drugs, cocaine, amphetamines, morphine, and heroin, have been legally sold in medicines such as cough mixtures, pain-relievers and 'tonics'. Many people became unwittingly and tragically addicted to drugs that doctors thought would cure them. Now these drugs are illegal, there is an additional problem, the user has no assurance of the quality of the drug.

The cost of addictive drugs to our society is immense. So much violence, anti-social behaviour, distress and disease, is caused by these chemicals. Those who seek to profit from the illegal trade rely on mass corruption of governments and business, organized crime, murders and robberies, to make their enormous profits. Their addicted victims usually resort to crime to sustain their needs. Few school playgrounds, street corners, clubs, bars or parties, are free of the menace of these dangerous, deadly substances.

And what of the future? Will society or human nature change so that people no longer crave dangerous drugs? Will scientists understand more of the complex workings of the brain so that the vice-like grip of addiction can be broken?

The prospects are good for new cures for genetic disease, cancer and many other serious illnesses. The biggest biology experiment in the world, mapping the human genome, may be an important help. More battles against infectious disease will be won, and deadly scourges wiped from the planet (sometimes, of course, to be replaced by completely new diseases). Useful and safer medicines will mean longer and healthier lives.

Ancient humans may not have known how medicines worked, but they certainly understood their power. Science and technology allows us to harness that power.

Drugs are an important part of your life. You have the knowledge to use them wisely, to use them well.

Index and Glossary

the world. Made from the chemical salicylic acid and first discovered in the bark of the white willow tree   38, **68–73**

asthma a type of allergy. A narrowing of the airways of the lungs which can be reversed, either naturally or by drug treatment. The airways of asthma sufferers also become hypersensitive to things like cold air, pollution and cell-destroying chemicals made by the body during the allergic reaction. Pollen and spores from fungi can trigger an attack, as can many substances in the home, especially the faeces of the house dust mite   56, **62–64**, 109

atom every form of matter in the world is made of atoms, which are the building blocks of any chemical element. Atoms join together to make molecules   8

axon a long strand of nerve cell cytoplasm that carries an electric signal   26, 27, 85, 89

AZT (azidothymidine) a drug used in the treatment of the HIV virus infection AIDS. Drug stops the HIV virus making DNA from RNA   31

b

bacteria microscopic single-cell forms of life that are a vital part of the balance of life. Some cause human illnesses   **12–20**, 29, **36–39**, 51, 56, 57, 114

basophil (bays-oh-fil) relative of the neutrophil white blood cell but much rarer in the blood. Basophils destroy bacteria but they also carry granules of cell-destroying chemicals which they can release during allergic reactions   **58–60**

BCG vaccine live vaccine of harmless bacteria closely related to the germ that causes tuberculosis. BCG stands for 'bacillus of Calmette and Guerin' the scientists who first isolated it   38

beer alcoholic drink usually made from fermented malt and flavoured with hops   90, 93

blood pressure force generated by heart beat and pressure of blood against vessels, especially arteries   80, 87, 92

blood vessel tube that carries blood. Arteries, veins and capillaries are all blood vessels   45, 58, 60, 73, 92

bone marrow soft centre of many bones where red blood cells, white blood cells and platelets are made   45, **48–50**, 52, 53

breathalyser instrument used to measure the amount of alcohol in the breath and hence how much a person has drunk   92

bug can mean a small insect, especially one that sucks blood or infests beds. Also used for microbes like bacteria or viruses and the diseases caused by them   **12–43**

butane gas easily liquefied gas used as a fuel and solvent. Household substance that is abused   85, 94

c

caffeine mild stimulant drug found in tea, coffee, cola nuts, cocoa and chocolate bars. Member of a group of chemicals called methylxanthines. Little evidence that moderate consumption of these stimulants is harmful to health or that they are seriously addictive   101, 106

cancer disease where the cells of the body go out of control. Cancer happens because the DNA plans of a cell are somehow changed. They may be damaged by chemicals or high-energy rays from radioactivity or the sun. They may be altered if mistakes are made when the DNA plans are copied inside the cell   **44–55**, 87, 103, 104, 112, 116

cannabinoids (can-a-bin-oids) active ingredients of the cannabis plant that affect nerve cells in the body   87

Cannabis sativa bushy plant that yields one of the most commonly used intoxicant drugs. Its powerful resin, compressed into blocks for smoking or adding to food, is known as hash, dope, pot or ganga. Its leaves are smoked as marijuana or grass   **85–87**, 111

carbohydrate (car-boh-high-drayt) chemical that provides source of energy for the cell   40

carbon dioxide (CO_2) colourless odourless gas produced during breathing or when organic (carbon-containing) chemicals are burned. Animal life takes in oxygen and breathes out a mixture of oxygen and carbon dioxide   90, 103

carbon monoxide (CO) a colourless odourless and poisonous gas produced when carbon-containing chemicals burn with insufficient air. The gas binds tightly to haemoglobin in the blood and stops it taking up oxygen. One of the dangerous gases found in cigarette smoke   103

cell microscopic building block of living creatures. Some creatures are made of just one cell (e.g. bacteria), humans are made of about 100 million. A cell is not just a bag of chemicals, it is a highly organized and very complicated entity. Each cell is like a miniature factory estate   7, 9, **44–55**

cell division process whereby one cell divides into two after copying its DNA plans   52

cell wall (in bacteria) bacteria are surrounded by a complicated semi-rigid cell wall that gives them their characteristic shape. This protects bacteria from adverse conditions. Some antibiotics damage the cell wall, causing bacteria to burst   12, 19, 38

Chain, Ernst (1906–1979) Anglo-Russian scientist who first purified penicillin and discovered ways of making enough to treat patients. Awarded Nobel Prize for Physiology or Medicine in 1945 along with Fleming and Florey   **15–17**

chemotherapy (keem-oh-therapy) treatment of a disease with chemical substances, e.g. the cell-damaging chemicals used to treat cancer   18, 38, **48–54**

chickenpox common herpes virus infection,

usually in children. Causes itchy spots and fever. Patient normally gets better within a few days 24

chloroform (klor-oh-form) a dense liquid used as a solvent and as an anaesthetic 79, 80, 85

chromosomes (krohm-oh-sohms) thin strands of DNA (see below) found in the nuclei of cells. Human beings have 46 chromosomes 10, 12, 20, 22, 30, 31, 52

chronic infection infection that goes on for a long time and is difficult to cure 29, 42

cigarette small cylinder of finely cut tobacco or other drug, rolled in paper for smoking 46, 62, **102–105**, 114

cinchonine (chin-choh-neen) fever-reducing drug extracted from the bark of the Peruvian fever tree. Used as a treatment for malaria 42

citrus fruit from a group of plants that bear acid-tasting fruits. The fruit is in fact a type of berry enclosed in a thick leathery rind. Lemons, oranges, grapefruit, limes, mandarins and tangerines are common citrus fruits 47

'Coca Cola' drink of world-wide popularity first invented in 1886, when it contained very different ingredients from the present day. The 1886 version contained both alcohol and cocaine, neither of which are found in this drink today 106

coca shrub bushy shrub, *Ethroxylon coca*, whose leaves are made into a paste and chewed. Chewing the leaves causes numbness of the mouth and has a stimulant effect, reducing fatigue and altitude sickness. The addictive stimulant cocaine is purified from coca leaves 82, **106–108**

cocaine highly addictive stimulant extracted from the leaves of the coca shrub. The white powder is usually laid out on a flat surface in lines and sniffed through a straw up the nose. It can also be dissolved in water and injected. Crack or freebase is the most powerful and addictive form. Users of cocaine experience brief euphoria, increased alertness, energy, physical power and mental clarity, followed by a crash into severe depression 82, 87, **105–108**, 110, 115

coffee evergreen shrub with cherry-like fruits containing two seeds (coffee beans) which are rich in caffeine and used to make the popular mildly stimulating drink 85, 100, 101

cola evergreen tree native to tropical Africa that produces nuts high in caffeine that is released when the nuts are chewed 100, 101, 106

cold sore caused by a herpes virus infection. Blister-like rash affecting lips and skin around the mouth, the lining of the mouth and the tongue. Between attacks the virus lies dormant **24–27**

colony (of bacteria) a clone of bacterial cells that grows from the repeated division of one bacterium into millions and makes a raised disc-shape on a solid nutrient surface 13, 14

coma state of prolonged unconsciousness 42

condom contraceptive sheath worn on penis during sexual intercourse 31

cowpox a virus infection of cows' udders that can be transmitted to the hands of humans who milk the cow. Caused by a pox virus very similar to smallpox, and capable of stimulating immunity to smallpox when used as a vaccine 33, 34

'crack' name given to highly addictive freebase cocaine. Made by chemical treatment that frees the cocaine base from hydrochloride. Small 'rocks' are smoked in a pipe or heated in tin foil 108

cromolyn (Intal) drug used to treat asthma. Originally purified from the khellin plant used by Egyptians to treat breathing disorders 64

curare extract of S. American plants, used to make poisoned arrows. Active ingredient, tubocurarine, used to paralyse muscles during surgery 81

cytokines (sigh-toh-kyne) messenger molecules made by defender cells that help control the fight against infection 55

cytoplasm (sigh-toh-plas-m) a clear liquid that surrounds everything in the cell. It has many very fine protein fibres that give the cell its shape and allow it to move, especially when one cell divides into two 9, 12, 22, 23

d **dander** underfur of animals that can provoke an allergic reaction 56, 57

Davy, Sir Humphrey (1778–1829) British chemist. Discovered many elements including potassium, sodium, barium and calcium. Also invented the miner's safety lamp. The first person to suggest that nitrous oxide (laughing gas) might be useful in surgical operations 78

defence system (see immune system)

defender cells cells of the body's defence system that work together to fight infections. Lymphocytes, macrophages, and neutrophils are all defender cells 13, 19, 27, 37, 38, 54, **56–65**

depression a mental condition in which the person suffers from low mood, low self-esteem and despondence 87, 96, 107

developed world mainly used to describe the wealthier countries of Europe, America and Asia 39

diabetes (digh-a-beet-ees) a common disease in which the pancreas fails to produce enough of the hormone insulin. Insulin normally controls sugar metabolism and in diabetics levels of sugar in the blood can become dangerously high 50, 112

diesel exhaust particles less than one-thousandth of a millimetre in diameter, these particles are found in polluted atmospheres. They are thought to make allergies worse 56, 57

diphtheria (dif-theer-ee-ah) infectious disease

caused by the bacterium *Corynebacterium diphtheriae*. Disease begins with a sore throat and is followed by a general illness and a swelling of the throat. The bacteria produce a powerful toxin that interferes with the cells' protein-making machinery. As little as one-hundredth of a milligram will kill someone. Now rare because of vaccination with the inactivated toxin   36

disease any unhealthy condition of the body or mind. An illness or sickness   **13–65**, 70, 71, 73, 87, 97, 98, 104, 112, 114, 116

distillation to remove dissolved impurities or separate two liquids by heating and then cooling the vapour. Whisky, brandy and other strong alcoholic drinks made by distillation   90

DNA (deoxyribonucleic acid) the chemical that contains an organism's genetic information. Long thread-like molecule found in the nucleus of cells in strands called chromosomes. The DNA in each human cell contains recipes for making between 50,000 and 100,000 different types of protein. A DNA recipe for a single protein is called a gene   **10–12**, **19–25**, 30, 31, **44–47**, 55, 104

dope term sometimes used for cannabis but also for other abused drugs   86

drug any chemical that can change the way a living creature functions. Drugs can be simple chemicals or complicated ones. They can be inhaled, swallowed, absorbed through the body surface, or injected. But all drugs work in the same way. They get very close to natural chemicals in living cells and change what they do   **4–11**, **14–116**

e **'Ecstasy' 'E'** a 'designer drug' that is meant to be mildly hallucinogenic. MDMA (methylenedioxymethamphetamine) is its chemical name. Usually a tablet or capsule, often mixed with amphetamine sulphates, tranquillizers or LSD. Also known as 'Edwards', 'disco biscuits', 'disco burgers' and 'Dennis the Menace'   85, 109, 110

eczema (ek-si-mah) common inflammatory disorder of the skin   56, 64

endorphins natural opiates that affect the body's perception of pain. Endorphins are made in the brain at times of stress and danger. Acupuncture may work by stimulating the body to make endorphins. Also known as enkephalins   75, 99, 111

enzyme a protein molecule that acts as a catalyst allowing chemical changes in other molecules, without being altered itself. Enzymes carry out digestion of food; bacteria produce enzymes that destroy antibiotics; enzymes control many of the chemical reactions inside a cell   9, 20, 23, 24, 55, 71, 82

eosinophils (ee-oh-sin-oh-fils) rare relatives of the neutrophil defender cells. Made in the bone marrow and circulate in the blood. When the immune system is functioning normally, their vital job is to destroy the larvae of parasitic worms that invade the tissues. In allergy their powerful killing weapons damage human cells   58, 59, 63

ester a chemical compound made by condensation of an acid with an alcohol   90

ethanol proper chemical name for alcohol   90, 93

ether volatile inflammable liquid used as an anaesthetic   78, 80, 85

ethyl alcohol another name for ethanol   90

euphoria (yoo-for-ee-ah) feeling of well-being based on over-confidence or over-optimism   87, 98, 106, 108, 109

f **faeces** waste material discharged from the rectum. Made from bacteria, food residue, water and other waste products   62, 65

Faraday, Michael (1791–1867) British scientist who made a number of important discoveries in the physical sciences, including electromagnetism, electrolysis, condensation of gases and studies on benzene and steel. Faraday took great trouble to make the latest scientific discoveries intelligible to the general public and started the Royal Institution Christmas Lectures for children   78

fats chemicals that can generate energy. One type of fat molecules, called lipids, also make the important membranes inside and surrounding cells. Essential part of our diet, turned by the body into energy stores. Fat cells beneath the skin act as a shock absorber and keep you warm   22, 92

ferment to break down a chemical with an enzyme. Typically carried out by enzymes in yeast that break down sugars to produce alcohol   90

flatworm flattened worm-like animal. Flatworms, like tapeworms and flukes, are parasites and cause disease   40

Fleming, Sir Alexander (1881–1955) British bacteriologist who discovered the first antibiotic, penicillin, in 1928. The significance of his discovery was only fully appreciated some ten years later when Florey, Chain and their colleagues purified and then manufactured penicillin on a larger scale. Awarded Nobel prize in 1945 together with Florey and Chain   14, 15, 17

Florey, Sir Howard (1898–1968) Australian-born doctor who worked with Ernst Chain to purify and manufacture the first antibiotic, penicillin. Awarded Nobel prize with Chain and Fleming in 1945   **15–17**

'freebase' another name for crack (see crack)

Freud, Sigmund (1856–1939) Austrian pioneer of psychoanalysis   82, 106

fungus (*pl.* fungi) a primitive form of plant that obtains nutrients by secreting enzymes to break down food externally, or by becoming a parasite and absorbing food from its host. Fungi destroy crops and cause disease. Some are poisonous but others, like mushrooms, are edible   62, 88

g   **GABA** (gamma-aminobutyric acid) powerful neurotransmitter. Present at 25–40% of all gaps (synapses) between nerve cells. Alcohol increases the activity of GABA   111

ganga (gan-jah) another name for cannabis   86

gate control theory idea that nerve signals have to pass through a series of 'gates' where the strength of each signal is measured. If the signal is strong enough it will carry on up to the next gate and eventually to the brain   75

gene DNA recipe for a protein. Human DNA contains about 100,000 different genes   29, 55

genetic defect mistake in, abnormality of, or damage to, DNA. This can cause an abnormal protein(s) to be made, or the protein may not be made at all (see also mutation)   109, 112

genetic engineering manipulation of DNA and particular genes, e.g. inserting the DNA recipe (gene) for a human protein into a bacterium so that the bacterium then makes the human protein along with its own. Used to make useful drugs   29

genetic instructions (see DNA)

genetic material (see DNA)

germ microbe that can cause disease   12–43

glucocorticoids (gloo-coh-cort-ik-oids) hormones that control protein and carbohydrate metabolism, and are also natural suppressors of the immune system. Glucocorticoids, and substances that resemble them, are very powerful as anti-inflammatory drugs   64

glue sniffing form of solvent abuse that causes intoxication   94

gram unit of mass. 1000 grams in a kilogram   15, 28, 60, 88, 89, 93, 101

guanine one of the four chemicals (bases) that make up the DNA genetic code   10, 11, 25

h   **haemorrhage** (hem-or-ij) loss of blood   104

hallucination (hal-oo-sin-ay-shon) an illusion, a vision, an apparent perception of an external object not actually present   4, 74, 87, 95

hallucinogen (hal-oo-sin-oh-jen) drug that induces hallucinations   4, 74, 85, 88, 89

halothane a type of anaesthetic   80

hangover unpleasant after-effects of drinking too much alcohol   92

hash another name for cannabis   86

hay fever (medical name allergic rhinitis) most common form of allergic disease affecting up to 15% of the population. Allergy to substances such as pollen and animal fur causes runny nose, sneezing, watery eyes, etc.   56, 59, 60

heart attack occurs when part of the blood supply to the heart (coronary artery) becomes blocked by blood clot   94, 104

hemp popular name for the plant *Cannabis sativa*, source of rope fibre and bird seed, but best known for its intoxicant resin   86

hepatitis virus infection of the liver   29, 98

herbal remedy any drug based on a plant or plant extract   5, 64, 68

heroin one of the opiates derived from the opium of the poppy. Chemically produced by adding two acetyl groups to morphine. Originally sold as a cough medicine until its powerful addictive nature was discovered   85, **97–99**, 105, 115

herpes virus virus that causes diseases like chickenpox, cold sores and glandular fever **24–27**

heterosexual person who is attracted to people of the opposite sex   30, 31

histamine cell-destroying chemical released by mast cells and basophils which causes many of the symptoms of allergies like hay fever and asthma. Anti-histamine drugs block its action   60, 62

HIV Human Immunodeficiency Virus, the virus that causes the disease AIDS   30, 39

Hoffman, Felix (1868–1946) chemist who found that by subjecting salicylic acid to a chemical process called acetylation, he could make a better painkiller now known as aspirin   70

Hofmann, Albert (1906–) chemist at the Sandoz Drug company in Switzerland who made LSD-25 in 1943 and then, by accident, discovered its hallucinogenic properties   88

homosexual person who is attracted to people of his or her own sex   30, 31

hormones chemical messenger molecules that affect the body's cells and organs   49, 51, 52, 55, 63, 64, 105, 109, 111

hypnotic a drug that causes a state of sleep, calm and stupor   85

hypodermic syringe a syringe with a needle point that allows a drug to be given under the skin. Also used to inject drug into a muscle or vein   96

i   **IgE** (immunoglobulin E) a type of antibody produced by lymphocytes that binds to cells called basophils and mast cells. IgE is the antibody that triggers allergic reactions, although it is thought to be useful in the fight against the larger disease-causing parasites   **58–60**, 62

immune system collection of cells and organs in the body that work together to fight infections. The immune system is really made of two parts:

the innate system which acts as a first line of defence and prevents the majority of germs causing any trouble, and the adaptive system which makes a specific defence against each infection. The adaptive system can remember particular infections and prevent them causing disease later. The cells of the immune system (defender cells) include lymphocytes, macrophages, neutrophils, basophils and eosinophils. These produce antibodies, cytokines and other proteins that help fight infections   24, 27, 30, 31, 58, 64

immunoglobulin (imm-yoo-noh-glob-yoo-lin) another word for antibody   **58–60**, 62

immunotherapy (of allergy) treatment of allergy by giving sufferer repeated doses of the substance they are allergic to, so that the allergic reaction is eventually abolished. Only used in people with very bad allergies because it can lead, itself, to severe allergic reactions   64

inebriant (in-ee-bree-ant) drugs that make the user euphoric or drunk. The most common inebriant is alcohol although ether, butane and solvents have similar effects   85, **90–94**

infection invasion of the body by microbes that are capable of growing there and causing illness   **14–43**, 112

inflammation defence reaction of the body to invasion by microbes, the presence of a foreign body, or other tissue damage. Defender cells like lymphocytes, macrophages and neutrophils are attracted to the site. An inflammatory reaction is accompanied by local swelling and pain   63

influenza respiratory illness caused by influenza virus accompanied by chills, fever, headache and muscle pain. Each year epidemics of 'flu spread rapidly around the world   29, 31, 43, 73

inhaler delivery system that allows drugs to reach the lungs   **62–64**, 109

insecticide any chemical which kills insects   41, 43

insect small animal without a backbone that has head, thorax, abdomen, two antennae, three pairs of legs and two pairs of wings. Can carry disease, e.g. malaria and *Anopheles* mosquito   41

interferon a protein produced by cells that protects other cells nearby from infection by viruses. Part of the body's natural defence against viruses and other infections. Interferons are cytokines (see above)   28, 29, 55

intestines part of the digestive system that extends from the stomach to the anus   12, 49, 53, 92

intoxicant a drug that alters the mind. Hallucinogens, hypnotics, inebriants and stimulants are all intoxicants   **84–111**

Isaacs, Alick (1921–1967) co-discoverer, with Jean Lindemann, of interferons (see above)   28

Isoniazid drug used to treat tuberculosis   38

j   **Jenner, Edward (1749–1823)** English country doctor who discovered the process of vaccination. His experiments with cowpox in 1796 founded an area of medicine that eventually resulted in the development of safe, effective vaccines, and the complete eradication of smallpox from the world   33, 34, 37

k   **khellin** plant used by ancient Egyptians from which the asthma drug cromolyn (Intal) was originally purified   64

kidneys important organs that filter out harmful waste products from the blood, and keep the fluid balance of the body   92, 101, 104

killer cell a type of lymphocyte that recognizes and kills damaged or altered cells   31, 37, **54–55**

l   **laughing gas** another name for the gas nitrous oxide, used as an anaesthetic   78, 79

leukaemia cancer of the bone marrow which results in too many white blood cells   45, **48–53**

Lindemann, Jean (1924–) co-discoverer, with Alick Isaacs, of interferon (see above)   28

Liston, Robert (1794–1847) first British surgeon to use anaesthesia in 1846   79

liver important organ that is the body's chemical processing plant   29, 45, 82, 92

local anaesthetic drug that works by preventing nerve cells transmitting pain signals from a limited area   82

locus coeruleus (loh-cus see-roo-lee-us) part of the brain stem especially involved in emotions. Nerve cells of the locus coeruleus have long thin axons that reach for vast distances all over the brain, branching so that they touch as many as one-third of all other nerve cells   88, 89, 110

Long, Crawford (1815–1878) doctor from Georgia, USA who, in 1842, used ether for an operation. Failed to publicize his success   78

LSD lysergic acid 25. Powerful hallucinogen. Its properties were discovered accidentally by Albert Hofmann in 1943   85, 88, 89, 109, 111

lymph vessel (limf) vessel that carries watery fluid full of defender cells around the body. Part of lymphatic system   45

lymphocytes (lim-foh-sights) important cells of the immune system. Make antibodies and can become killer cells that attack and destroy altered, damaged or virus-infected cells   27, 37, 54, 58, 60

lysergic acid (lie-ser-jick acid) (see LSD)

m   **Ma huang** plant used to treat asthma in ancient China. Its active ingredient was ephidrene. Amphetamine is a synthetic substitute   109

macrophages important cells of the immune

system that engulf and destroy germs, help lymphocytes make antibodies and killer cells, and clear up debris in the body   27, 38

malaria infection caused by the organism *Plasmodium* and spread by the *Anopheles* mosquito. Kills one to two million people annually, most of them children. *Plasmodium* breeds in the female mosquito and infects humans while the female is sucking their blood. A typical symptom of malaria is a violent fever lasting 6–8 hours recurring every 2 or 3 days   **41–43**

mandrake root (*Mandragora officinarum*) one of the most famous plants to be used by witches and sorcerers. Contains substances that are hypnotic, hallucinogenic and soporific. Also causes vomiting and is a painkiller   78, 85

marijuana (mari-wah-nah) leaves from the plant *Cannabis sativa* smoked to obtain their intoxicating qualities   **86–87**

mast cells cells involved in allergic responses that release cell-destroying chemicals like histamine when triggered by allergen   **58–60**, 63

measles extremely contagious virus infection that is spread through the nose and mouth. Vaccination with a weakened form of the virus is effective   31, 36

membrane (of a cell) a double layer of fats called lipids. Rather similar to the flexible surface of a soap bubble. Very little can get through, only substances the cell really needs, no longer needs, or is sending to other cells   9, 22, 81, 87, 92

messenger strand (see RNA)

metastases (met-as-tas-sees) cancer cells that have spread from where the cancer first occurred to form tumours elsewhere in the body   45, 55

methylxanthines (mee-thile-zan-theens) mild stimulants, e.g. caffeine, found in drinks like tea, coffee and cola-flavoured drinks   101

microbe a living organism too small to be seen with the naked eye   **12–43**

microscopic too small to be visible without a microscope   29, 32, 90

milligram unit of mass, one-thousandth of a gram, one-millionth of a kilogram   93

millipede long-bodied animal without a backbone. Some millipedes can spread disease   41

mineral belonging to any of the classes of inorganic (without carbon) chemicals   4

mite small short-bodied arthropod with head and abdomen fused into a compact body. Usually has four pairs of walking legs. Faeces of the house dust mite can cause asthma   41, 62, 65

mitochondria (migh-toh-con-dree-ah) like miniature power stations that make, supply and store energy for the cell   9, 19

molecule two or more atoms combined by a chemical reaction. The smallest physical unit of any compound   8, 9, 74, 111

morphine (mor-feen) powerful hypnotic drug made from the opium poppy   **74–76**, 96, 97, 99, 106, 115

Morton, William (1819–1868) dentist who was the first to demonstrate the success of ether anaesthesia. Fought with Horace Wells and Crawford Long for recognition of his discovery and died a pauper   79

mould furry growth of fungi on animal or vegetable matter   **14–18**

mucus (mew-cus) thick fluid secreted by parts of the body cavity that are open to the exterior   60

muscle relaxant drug used to relax muscles during operations under general anaesthesia 80, 81

mutation (mew-tay-shon) change or alteration. Usually refers to a change/mistake/damage to DNA of a living creature which can be inherited and thus causes variation in the species. Mutations in the DNA of bacteria can make them resistant to antibiotics. Mutations are the reason why life on earth has evolved from minute single-cell creatures   20

n   **naphthalene** (nap-tha-leen) a white waxy solid obtained from distilling coal tar. Important starting material for many dyes and plastics   72

nebulizer machine that allows high doses of drug to be delivered to the lungs, e.g. in asthma. The drug and other liquid is placed in a chamber and compressed gas passed through. Aerosol then breathed in through a mask   63

Neimann, Albert (19th century) chemist who extracted cocaine from coca leaves in 1860   82, 107

Neolithic (nee-oh-lith-ik) Stone-age early humans who used polished stone weapons and tools   96

nerve cell cell of the nervous system, also known as a neurone. Processes and transmits information around the body. Typically, a nerve cell has a cell body with a nucleus surrounded by cytoplasm, an axon which transmits information, and dendrites which receive information   24, 26, 30, 60, 66, 71, 81, 82, 85, 87, 89, 92, 103, 105, 110

neurotransmitter (new-roh-trans-mitter) messenger chemical made by nerve cells that allows them to send their signals to other nerve cells. Nerve cells have long strands of cytoplasm called axons that end in nerve endings with one thousand or more branches. Each branch may connect with a different nerve cell, either directly on the main nerve cell body, or more usually, through up to ten thousand projections of cytoplasm called dendrites. Signals are rapidly carried along axons by a change in electrical charge. But when a signal comes to a nerve ending it is transmitted across the gap (synapse) to the next nerve cell by the rapid release of neurotransmitters. The neurotransmitters briefly stick to 'receptor' molecules on the adjacent nerve cell, as a key fits in a lock. The neurotransmitters can either pass on an 'excitatory' or 'inhibitory' message. Intoxicant drugs work by changing the way neurotransmitters work   85, 94, 110, 111

neutrophils (new-troh-fils) white blood cells made in the bone marrow. Important cells in defence against infection because they ingest and destroy invading microbes. Produced at a rate of 80 million a minute and only live 2–3 days. 70% of all human white blood cells are neutrophils   58

nicotine addictive stimulant found in the plant *Nicotiana tabacum*, and in tobacco made from the leaves of that plant   85, **102–105**, 111

nitrogen mustard gas (also known as sulphur mustard) chemical warfare gas introduced by the German Army near Ypres in 1917. During the Second World War, scientists discovered that it killed rapidly multiplying cells and began to use it, and similar chemicals, to treat cancer   48

nitrous oxide gas with a slightly sweet odour that is used as a short-acting anaesthetic. Also known as laughing gas   78, 80

Nobel Prize prize awarded each year from a trust fund established in the will of Alfred Nobel. Given to people who have done important work in physics, chemistry, physiology or medicine, literature and peace   17

nocioceptor (noh-see-oh-sept-or) a nerve ending that detects pain   66, 67

noradrenaline (nor-ad-ren-a-lin) hormone produced by adrenal glands. Also a neurotransmitter; other name norepinephrine   105, 111

nucleic acid (new-clay-ik acid) another name for RNA and DNA (see RNA and DNA)

nucleus part of the cell which contains the DNA chemical instructions that control it   9, 22

O   **opiate** (oh-pee-ayt) chemical that binds to opiate receptors (see below)   **75–77**, 98, 99, 111

opiate receptors receptors in the brain that bind opiates like morphine and heroin. Also bind the body's natural opiates, endorphins   75, 99

opium extract of poppy plant, *Papaver somniferum*, used as a drug. Milky juice of unripe poppy seed capsules. Contains powerful chemicals including morphine   74, 78, 85, 96, 97

p   **pain** unpleasant sensation. Stimulation of nerve endings by heat, pressure or tissue damage. Pain receptors (nocioceptors) found over most of the body surface and inside   **66–76**, 82, 96

pain receptors (see nocioceptors)

painkiller drug that relieves pain   **66–83**, 96, 99

paracetamol (para-set-a-mol) commonly used painkiller   **72–74**

parasite animal or plant living in or on another and taking nutrients from it   22, **41–43**, 58

pasteurization (pas-ture-ize-ay-shon) process of heating food to destroy microbes. Named after the famous French scientist Louis Pasteur (1822–1895) who also discovered that disease was caused by germs   39

penicillin first antibiotic drug. Initially isolated from the mould *Penicillium notatum* by Alexander Fleming in 1928   **14–19**, 38

penicillinase penicillin-destroying enzyme produced by resistant bacteria   20

periwinkle evergreen shrub from which the anti-cancer drug vincristine was extracted   49, 50

Peruvian fever tree (*Cinchona*) bark from this tree used by Incas for treating and curing fevers. Contains substances such as quinine that are important in treating malaria   42

phenacetin (fen-a-set-in) painkilling drug now replaced by paracetamol   73

pill small ball, flat round tablet, or bullet-shaped form of medicine that can be swallowed   20, 60, 70, 107, 110

placebo (pla-see-boh) dummy pill or medicine given as a control in trials of new drugs   113

plasmid a small circular molecule of DNA found in bacteria that replicates independently of the bacterial chromosome. Plasmids often contain genes conferring resistance to antibiotics   20

Plasmodium (plas-moh-dee-um) (see malaria)   42, 43

platelets fragments of cells found in blood and important in blood clotting   71

polio virus that causes paralysis. Vaccination programme stands a good chance of clearing the world of this disease   36

pollen male sex cells of seed plants. Pollens from some grasses and trees provoke allergies   58, 60, 62, 64

'pot' another name for cannabis   86

potion dose or draught of liquid medicine   5, 112

pre-med drug given to patients just before an operation to relax them and give relief from pain   80

prostaglandins (pros-ta-gland-ins) hormone-like substances that are produced in many tissues of the body and circulate in the blood. Involved in inflammation. Painkillers like aspirin and paracetamol stop prostaglandins working   71, 73, 74

proteins complicated chemicals that make cells the size and shape they are. Proteins allow a cell to move. Proteins called enzymes help the cell carry out all of its chemical reactions   9, **19–23**, 28, 52, 60, 75, 81, 94, 99

pulse pressure wave caused by the blood being pumped from the left ventricle of the heart into the blood vessels   80, 87

'purple heart' name given to amphetamines abused by American soldiers, especially in the Vietnam War   109

q **quinine** drug used in the prevention, and sometimes treatment, of malaria. Originally derived from the bark of the Peruvian fever tree   42

r **radioactivity** decay of some atoms that leads to the emission of particles and energy. Materials that contain these unstable atoms are called radioactive. There are three main types of radioactive particle: alpha, beta and gamma. Radioactivity, particularly gamma rays, can damage living cells, especially their DNA   44

Rastafarians members of a religious movement that started in the West Indies. Believe that the former emperor of Ethiopia, Ras Tafari (Haile Selassie), was their Messiah and Ethiopia their promised land. Normally wear their hair in dreadlocks   86

receptor embedded in cell membranes, receptor molecules receive messages from other cells. When a messenger molecule fits into its receptor, chemical signals are generated in the cell and travel rapidly to the nucleus command centre   60, 75, 82, 85, 87, 99, 111

ribosomes (rye-boh-sohms) like miniature factories that assemble proteins in cells. Some ribosomes float free, others are stuck on the walls of tiny tunnels which store and deliver proteins the cell has made to other cells nearby   9, 19

RNA (ribonucleic acid) the messenger strand that carries the DNA recipe for a protein to a ribosome. When a cell needs to make a protein, the segment of DNA that is the recipe for that protein starts to unwind from its chromosome. An RNA copy of the recipe is made from one of the two DNA strands. It is made of adenine, guanine and cytosine, just like DNA, but uracil replaces thymine. The RNA copy strand floats off out of the nucleus and finds a ribosome. Proteins are made from building blocks called amino acids. The RNA instructs the ribosomes to join up the amino acids in a precise order which is different for each protein   **20–24**, 30, 31

roundworms cylindrical worms that can become human and animal parasites and cause disease   40

s **salicylic acid** (sal-i-sil-ik acid) white powdery crystals that form the active ingredient of willow bark. Named from the Latin for willow (*Salix*).

Aspirin was made by a chemical change (acetylation) to salicylic acid   69, 70

schizophrenia (skitz-oh-freen-ee-ah) a major mental illness with altered thoughts and perceptions, including loss of contact with reality. It may cause a change of personality   108

sensory nerves nerves that carry signals concerning touch, pain, heat, cold, etc. towards the central nervous system   89

serotonin (seer-oh-toh-nin) a type of neuro-transmitter (see above). The hallucinogenic drug LSD is thought to resemble serotonin   111

shaman (shay-man) priest or witch doctor   4, 84

smallpox highly contagious virus illness that probably affected 80% of the population of Europe in the Middle Ages and killed about 20%. Those who recovered were disfigured for life. Caused by a pox virus. First disease to be eradicated from the world by vaccination   **32–36**

smooth muscles also called involuntary muscles. Not under conscious control of brain, but move automatically, e.g. in the digestive system   101

solvent the part of a chemical solution containing a dissolved substance. Also describes organic substances used to make paints and glues that are sometimes abused as inebriants   85, 94, 95

spinal cord part of the nervous system protected by the vertebral column. Essentially a long, thick cable formed of thousands of nerves   53, 66, 67, 77, 110

spirits (alcoholic) liquids obtained by distillation of purified alcohol. Whisky, brandy, gin and rum are all spirits   90, 93

spores reproductive cells of plants. Can provoke allergies in humans   56, 57, 62

stimulant drug such as amphetamine and cocaine that abnormally inflates a person's mood so that they feel powerful and fearless. Stimulants cause feelings of wakefulness, alertness, elation and an increased capacity to concentrate. Some can be highly addictive   85, **100–109**

Stone, Rev Edward (18th century) a country vicar who thought that willow bark might be useful for treating fevers. His experiments were the beginning of the story of aspirin   68, 69

streptomycetes (strep-toh-migh-seets) bacteria that produce antibiotics like tetracyclin, erythromycin and neomycin   18

streptomycin (strep-toh-migh-sin) important antibiotic produced by streptomycete bacteria   38

'sulph' amphetamine (see above)

synapse (sigh-naps) gap between nerve cells that is bridged by neurotransmitters (see above)   111

t **tannins** derived from various plants especially tea.

Give drinks a bitter and astringent taste   90

tapeworm intestinal parasite with suckers on the head for attaching to the insides of the host intestine. Absorbs digested food from intestine   40

tar liquid formed when organic compounds like coal, and some chemicals in tobacco, are burnt in the absence of air. Tiny particles of tar in cigarette smoke contain many cancer-causing and irritating chemicals. Less than one-thousandth of a milli-metre across, they become lodged in the tiny alveoli of the lung. Coal tar contains substances important to the chemical industry and was the starting point of the discovery of several drugs including paracetamol   87, 103, 104

tea drink made from the cured leaves of a small evergreen shrub (*Camellia sinensis*) grown in Japan, India, China and Sri Lanka   85, 100, 101

tetanus infectious disease caused by the bacterium *Clostridium tetani*. Bacterium makes a powerful toxin that causes paralysis. Vaccine is a form of the toxin that stimulates the body to make antibodies to neutralize the real toxin   36

thiopentone type of anaesthetic   80

Third world under-developed countries of Latin America, Africa and Asia   114

ticks arthropods, related to spiders. Spread infection as they feed on humans. Cause diseases such as scabies, Lyme disease and rickettsias   41

tobacco dried leaves of *Nicotiana tabacum*   87, 102, 103, 114

tolerance (to drugs) a part of the addictive state (see above) where progressively larger doses of drug are required to give the same effect   98, 105

tranquillizer drug taken to relieve anxiety and agitation without inducing marked sleepiness   85, 111

tuberculosis (tew-ber-cew-loh-sis) (TB) infectious disease caused by bacterium *Mycobacterium tuberculosis*. A progressive disease with loss of weight, coughing, and tiredness. Antibiotic treatments began to work in the 1950s, but antibiotic-resistant strains are now causing problems   36, 38, 39

tumour usually refers to lump of cancerous cells   45, 49, 54

u **unit** (of alcohol) a glass of wine, a measure of spirits or a half-pint/quarter-litre of beer all contain 8–12 grams of pure ethanol – one unit   92, 93

v **vaccination** (vax-in-ay-shon) first started by Edward Jenner (see above) who injected people with cowpox to prevent them getting smallpox. Name comes from *vaccus*, the Latin for cow. Way of controlling disease by making people immune to a virus, bacteria or bacterial toxin. Smallpox has been eradicated because of vaccination and many important diseases are now controlled   32–37, 39, 43, 54, 112

vaccine (vax-een) suspension of microbes, part or product of microbes, used for vaccination (see above)   34–37

Vane, Sir John (1927–) British scientist who discovered how aspirin worked   71

vector a living creature that spreads infection from one organism to another   41

vincristine (vin-cris-teen) drug used to treat leukaemia, first purified from periwinkle   50, 52

virus very small infectious particle. When outside a cell viruses are lifeless, inert particles made of nucleic acid, fats and proteins. When they invade a cell, they take over the cell, making more viruses, usually destroying the cell in the process. Many important human diseases are caused by viruses   22–36, 114

vitamin organic substance present in minute quantities in natural foods. Essential for health   12, 47

w **Wells, Horace** (1815–1848) dentist from Connecticut who used laughing gas for anaesthesia in 1844. Discredited when a patient cried out during a demonstration. His former pupil William Morton (see above) successfully demonstrated ether anaesthesia in 1846. Morton and Wells fought for recognition. Wells eventually committed suicide   79

white blood cells (see defender cells)

whooping cough lung infection caused by the bacterium *Bordatella pertussis*. Vaccine of heat-killed bacteria protects children   36

willow a family of deciduous trees and shrubs. The bark of the white willow contains salicylic acid from which aspirin is derived   68, 69

withdrawal symptoms suffered by an addict when their drug is no longer available. Addict may become severely depressed, over-excited, hypersensitive, confused and experience sweats, nausea, and hallucinations   98, 105

x **X-ray screening** X-rays are invisible rays with a shorter wavelength than light. Useful for detecting disease because different parts of the body absorb rays to a different extent, but enough pass through to register on a photographic plate   39

y **Young Simpson, Sir James** (1811–1870) Professor of Midwifery in Edinburgh who, in 1847, discovered that chloroform was a useful anaesthetic after experimenting on himself and his friends   79

The authors wish to thank the following people who helped in the production of this book:

Editor **Susan Dickinson**

For Portland Press
Sophie Caygill, Adam Marshall and **Rhonda Oliver**

For reference and useful criticism
Wellcome Institute Library
Dr **John Champness**, Wellcome Laboratories, Beckenham
Ms **Donna Goodal**, Bayer plc, Newbury
Martin Jarvis, Institute of Psychiatry, London
Dr **Helena Scott**, Hammersmith Hospital, London
Dr **Judith Kingston**, St Bartholomew's Hospital, London
Dr **Simon Page**, General Practitioner, Bristol
Dr **Hilary Thomas**, Hammersmith Hospital, London
Professor **Sir John Vane**, William Harvey Institute, London

Tony Geddes

and Dr **Chris Pogson**
for the original idea